BREAKING THE COST BARRIER

BREAKING THE COST BARRIER

A Proven Approach to Managing and Implementing Lean Manufacturing

STEPHEN A. RUFFA
MICHAEL J. PEROZZIELLO

JOHN WILEY & SONS, INC.

New York • Chichester • Weinheim • Brisbane • Singapore • Toronto

ISBN: 0-471-381365

Printed in the United States of America

10 9 8 7 6 5 4 3 2 1

Our appreciation goes to the many people
who helped us gain the insight to write this book.
We also thank our literary agent Reid Boates,
Jeanne Glasser at John Wiley & Sons, and
those on her team who helped to make this book a reality.
We are especially grateful to our families
for their understanding and support during
the long hours we spent throughout this project:

To my wife, Staci,
my children, Adam and Emily, and
my parents for their encouragement.
S.A.R.

To my wife Susie,
my son, Stephen, and
my family.
M.J.P.

Contents

BREAKING THE COST BARRIER

Introduction

The manner in which aircraft are manufactured, one that took almost a century to evolve, has suddenly become obsolete. No longer is the customer's single focus on pushing the limits of technology to fly ever farther, faster, and higher. In an environment plagued by downturns in defense spending and increased competition, cost has now become a primary driver. Suddenly, dramatic transformation is necessary. There is no choice; the customer can no longer afford the practices of the past.

Many have acknowledged that it will take a drastic change in approach to make this shift. Yet, they have also become increasingly aware that this change may not come easily.

Over the past several years, this industry had put into place an array of Japanese tools and practices intended to drive down costs. Known broadly as "lean production,"[1] they have been adopted in the hope of producing savings similar to those touted by the Toyota Motor Company. However, those who have applied them have more often than not found little improvement. Very rarely have these techniques yielded any-thing near what they should have accomplished.

Our dealings with this industry have shown us that, de-spite the fact that almost every company with which we have worked has applied some of these tools and practices, each has implemented them quite differently. Their results have varied just as much. It became apparent to us that there does not exist a strong understanding of how best to implement lean production in an environment that is different from that in the Japanese automotive industry. While much is available on *what* should be done in this area, little is written on *how* these techniques should best be put into place.

Suddenly, we found ourselves in a position to help.

After spending well over a decade working with the aircraft industry in the design, testing, manufacturing, and servicing of a wide variety of military aircraft, we grew increasingly frus-trated with companies' efforts to contain costs. We watched factory after factory struggle with daily problems that disrupted their operations, distracted the workforce, and drove up the cost of their products. We searched for the means to take a closer look, to learn from this industry's hard-earned lessons, and to develop an approach for attacking this barrier.

We knew that the answer would likely be found in the re-sults of this industry's efforts. We could see that, when viewed individually, each case could offer very little insight. How-ever, if they could be assembled together, allowing them to be analyzed as a group, much more could be learned. Thus, while wide-ranging experimentation has not produced the results that were originally intended, these results still offer tremen-dous value. With the comprehensive data from these ef-forts—from success as well as from failure—a solid path to future successes could be developed.

Since competition had precluded most from sharing these lessons, no one company had sufficient access to gain the clear picture needed of the overall path to improvement.

How could the individual pieces of this puzzle be brought together to provide the answers so critical to advancing the industry to the next level of cost performance?

It was our unique opportunity to pull together this information that paved the way for this book.

The U.S. Department of Defense set the stage for this opportunity when it made affordability a predominate consideration for its next tactical aircraft, the Joint Strike Fighter. With no other new programs in sight, competitors were faced with an urgent need to address this consideration. As they searched for the means, we brought forward our concept of consolidating this rich field of fragmented lessons. The Joint Strike Fighter organization agreed to sponsor our study.

Much like ourselves, many across the industry saw this as a key opportunity. Seventeen of the most prestigious companies producing military and commercial aircraft, engines, and avionics enthusiastically invited us and our team of specialists to weeklong visits of their facilities. They permitted us to look at them as insiders, allowing us to roam unescorted, viewing sensitive and even proprietary information. We had been given the unprecedented degree of access that was essential to combining the lessons of an industry—the opportunity that many observers never believed we could be granted.

This book is written to tell our story.

This book is intended for those who are interested in understanding how to apply lean production in an environment other than that of the Japanese automotive industry. It is written to show why so many who have come to believe in the tools and practices of this philosophy have not yet found the path to implementing them. It demonstrates that, in an industry where serious impediments exist, the mere understanding of these techniques is not enough. Instead, we identify the core focus of this concept, and show how all of its tools and practices must work together to achieve this intent.

Breaking the Cost Barrier demonstrates that each of the

major tools and practices associated with lean production is aimed at addressing one underlying factory issue: production variation. Production variation manifests itself in a number of forms, including poor quality, part shortages, underutilized equipment, and unpredictable production schedules. Its impact extends across all aspects of production, from the ordering of supplies to the assembly of the final products. Our study demonstrated that only by adopting an approach that targets this variation can large reductions in cost be achieved. In fact, after applying a specific framework to progressively address the sources of factory variation, the level of factory-wide improvement in such key metrics as manufacturing cycle time, inventory, and cycle time variation is astonishing—up to 67%, 80%, and 60%, respectively. The overall production cost savings is even more remarkable—up to 25%. Moreover, we repeatedly saw that this can be achieved with little capital investment *within months of implementation.*

This book shows that dramatic improvement is possible despite the existence of external constraints that cannot be changed. Adopting a focus on what *can* be changed—those processes internal to the organization—can lead to dramatic improvement.

Breaking the Cost Barrier provides a historical perspective of the reasons for the aircraft industry's current condition. It illustrates how such tools as Manufacturing Resource Planning (MRP II), Just-in-Time (JIT), kanban, cellular manufacturing, Statistical Process Control (SPC), and even advanced design practices have been applied, and the specific lessons learned from this industry's experimentation in each of these areas. This book draws from specific factory cases to drive home these lessons.

The results of our industry study received great praise from across this industry, academia, and the government. In fact, it was because of this overwhelming response that this book was written. While the initial report did highlight major findings, it merely served to whet the appetites of many, including ourselves. Ever since these findings were first released, the demand for further detail has steadily grown. At the same time, our continuing efforts led us to a deeper understanding of the framework that we had originally observed. This has further helped us understand the uniqueness of our knowledge base, as well as our responsibility to spread the word.

A number of highly credible organizations have embraced the conclusions of this initial study. First and foremost, the Naval Air Systems Command provided the support to make this study possible. The Department of Defense's Joint Strike Fighter program served as the sponsor of this study, and has acknowledged that "in order for the Joint Strike Fighter program to achieve its aggressive affordability goals, improvements in the manufacturing area such as those addressed in this [study] must be embraced by the industry."[2] Managers across the aerospace industry have offered praise, using such words as *insightful* and *revolutionary* in their descriptions.

We recognize that our findings do not represent the end to this story. Much as the efforts of others before us led to the development of components that make up our framework to improvement, it is likely that our findings comprise only one aspect of a much larger model. Since we limited our focus to what could be achieved despite the substantial hurdles faced by this industry, even the large savings potential that we cite will likely represent only a fraction of what is ultimately possible. It is entirely possible that others after ourselves will find that our results represent only the foundation of a much larger framework for advancement.

Finally, these lessons should not be left solely for the purposes of the aerospace industry. Because of the range of products and complexities of the facilities that were used to identify our results, the same approaches are likely to be applicable to a number of other industries. We hope that our observations will stimulate the next wave of broad-based experimentation, one that can be met with much more consistent success for a range of industries around the country.

The views expressed in this book are those of the authors and do not reflect the official policy or position of the Department of Defense or the U.S. Government.

CHAPTER 1

An Opportunity for Advancement

It has been half a century since a team of contractor and government engineers gathered in the Muroc, California, desert to usher aviation into a new era. This group set out to do what no one was certain was possible—to break through a seemingly impenetrable wall, one that no one was sure could be crossed. On October 14, 1947, their efforts became historic when, at the controls of the Bell X-1 aircraft, pilot Chuck Yeager broke the sound barrier.

For years many had attempted this feat, but to no avail. Failure led to research, and yet again to failure, until many began to believe that this barrier truly could not be crossed. Through this, advancements were made in such key areas as transonic aerodynamics and high-speed propulsion, yet a seemingly insurmountable problem remained. As an aircraft approached the speed of sound, it began shaking violently, ultimately losing controllability. More than one pilot was reported to have lost his aircraft—and his life—because of this.[1]

If supersonic flight were ever to become a reality, this phenomenon had to be understood, and ultimately overcome.

While the goal was clear, it proved to be exceedingly difficult to achieve. Traditional laboratory methods could not be used to lend assistance; wind tunnels were completely useless in duplicating this phenomenon. Flight-testing was the only way to proceed, yet it was exceedingly expensive and hazardous.

So this team of military, government, and contractor personnel joined forces in search of the answer. After numerous experiments, Yeager and this joint government-industry team had noticed that, with a traditional aircraft configuration, the nose of the aircraft began to rise as the speed of sound was approached. Shock waves forming on the horizontal tail caused the aircraft to lose its controllability. As the team worked through the problem, they discovered what would be the one remaining barrier to breaking the speed of sound:[2]

> The big thing that came out of the whole program was that we found out, in order to control the airplane through Mach 1, we had to have a flying tail on it. That was the first time we had experimented with this flying tail on the X-1. That really was the answer to flying at supersonic speed.
> . . . And it took the British and the French and the Soviet Union five years to find out that little trick that we found out with the X-1. It gave us a quantum jump on the rest of the world in aerodynamics.
> —Brigadier General Charles E. "Chuck" Yeager[3]

With this final step, a seemingly impossible hurdle forever vanished. A new window was opened, one that allowed for rapid new advancement in flight performance. Shifting from hinged elevators to a fully movable tail or "flying tail" made controllability at supersonic speeds possible. This one step changed everything; with air travel no longer relegated to subsonic speeds, aircraft could now move forward to seemingly limitless records of performance. Each generation of aircraft was better, faster, more maneuverable than the last. It appeared that nothing could halt this pace of development.

Yet, once again the aircraft industry has reached an apparently insurmountable barrier, one that must be overcome if advancement can continue its rapid pace. This time it cannot be accomplished through flight-testing, nor with a breakthrough in any of the other traditional disciplines of aircraft design. This new barrier is not one of performance; this time, it is one of cost.

The aircraft industry suffers from an apparent dichotomy in its capabilities. It continues to demonstrate an awesome ability to press the limits of technology. Its product quality is inarguably high; tremendous care is taken throughout design and production to ensure the utmost levels of flight safety. The industry has a long history of resilience, snapping back from repeated swings in demand, from financial downturns, changes in customer focus, and wars. It has adapted well to an ever-increasing level of regulation and oversight. Yet despite its achievements, this industry has not been able to stem its dramatic rate of cost escalation.

For over a decade, industry leaders have seen that this trend cannot be sustained.[4] Now it appears that the crisis is closer than anyone might have envisioned—even the next generation of military aircraft may not be affordable. The Defense Department will no longer pay as it has in the past to reach the next level of performance. Increased technology is no longer enough; the customer now also demands efficiency.

The trend of cost escalation is depicted graphically in Figure 1.1, in which it can be seen that the unit price has increased dramatically for each new generation of military aircraft.[5] It could clearly be argued that, in the long term, some sort of action will have to occur—no fighter airplane is affordable at the cost of billions of dollars. However, this industry is now finding that even the short-term trends depicted here must be altered.

Even with such a dire situation at hand, we were astonished at how many think that a rapid, dramatic reduction in

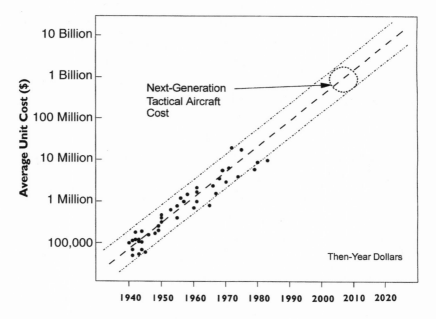

FIGURE 1.1 Cost escalation of tactical aircraft.

the cost of manufacturing is not possible. There seems to be a pervasive belief that the barriers are too daunting, that too much is out of the control of an individual facility. It was necessary to set aside this belief in order to cross the sound barrier. The industry must again rise to the challenge; it must find a way to overcome this barrier or face dire consequences. A new approach to manufacturing must be found.

It is the premise of this book that the magnitude of these barriers has been largely overstated. This is not to say that substantial barriers do not exist; the aircraft industry, like many others, must operate within an environment that is far from optimal. Wide swings in product demand, government regulation and oversight, and a strong dependence on military sales all affect its results. In addition, the unyielding demands for ever-increasing performance limit the trade-offs that could otherwise enhance design producibility, forcing factories to press manufacturing processes to their limits. Yet, in spite of this, we found countless examples of localized successes across this industry. By combining the lessons offered by these successes, we will show that a path to a much larger potential for savings is readily achievable, *even within the constraints of this existing environment.*

Our observations will be shown to boil down to a seemingly basic principle for success, one whose potential to lead to rapid, dramatic cost savings has already been demonstrated. Through the use of a simple framework for implementation, this principle can consistently enable an organization to reach these goals with a minimum of investment. We have termed this overall effect *the principle of variation management.*

Variation management offers a powerful new twist to manufacturing cost improvement. Unlike many other methods on which much has been written, it does not represent a single tool focused on affecting an individual aspect of production. Instead, it acts as the enabler that allows these tools to become most effective. By forming the core of a well-planned program, this principle can lead to widespread improvements within a facility, despite seemingly large obstacles. Even with only a limited understanding of this approach, improvement initiatives have sometimes resulted in phenomenal accomplishments.

With this framework we will show that, despite the many adverse external factors that cannot be readily changed, substantial improvement is still possible by adopting a focus on those internal processes of the organization that *can* be changed. The power of this approach has been repeatedly validated by real-world successes, with dramatic reductions in inventories, manufacturing cycle times, and production variability—and ultimately in production costs.

Facilities that have grasped aspects of variation management have largely become masters of their destinies, demonstrating an ability to respond to internal problems or external environmental changes much more effectively than others within the same setting. As its name suggests, this new approach is based on the premise that the waste caused by factory variation is a key driver to manufacturing costs. We found that this variation extends across all aspects of production, from the ordering of supplies to the assembly of final products. In order to deal with its day-to-day effects, factories must constantly apply tremendous resources, yet usually find themselves making little progress in reducing the underlying causes.

Using this principle, dramatic improvements to efficiency and responsiveness are possible throughout the supply chain. Suppliers are able to effectively manage the effects of widespread unforecasted demands from their customers, minimizing or even eliminating the disruption that otherwise would impact their ability to respond. Facilities that integrate and assemble these components are able to manage their own operations more effectively, allowing them to better forecast their specific requirements such that they can more smoothly load both their own shops as well as those that supply materials. They can dramatically reduce the need for work-arounds, expediting, out-of-station work, and stock-outs that disrupt factory flow. As a result, all are able to improve their performances, gaining significantly greater productivity and efficiency.

By removing much of the guesswork in implementing improvements, organizations whose efforts have followed this framework are much less likely to stumble early in their attempts, lessening the risks of disruption to production operations. These early successes can rally the workforce, setting the stage for long-term improvement. As a result, those that use this principle are much more likely to take on new approaches, and when they do, they are much more likely to succeed.

With the path to making such broad improvements well within reach, why do so many fail to see it? The answer may lie in how they view themselves today. We found that even in a factory marked by crisis management, one in which an inordinate amount of time is expended in developing work-arounds to day-to-day factory disruptions, many do not see a compelling reason to change. We spoke to workers and managers alike, with many explaining to us how this merely represents "the way aircraft are built." Yet, it seemed to us that the waste we found should cause concern; it clearly plays an important part in driving up manufacturing costs.

The following examples will help make this point clear:

■ A factory produces a complex product, one that requires a large number of successive manufacturing steps to be performed across a number of production shops. Even with a formal production scheduling system in place, at each shop we found expedite tags stapled to almost every order. Because of this large number of expedites, the formal scheduling system is crippled. Without any real way to prioritize these jobs, the shops are forced to respond to orders almost on a first-come-first-served basis. In essence, all jobs have been ranked as equally important. Yet, we found from those who have placed these orders that a priority to their need does exist. In fact, only a small set of the items ordered using expedite tags are needed immediately to prevent work stoppages. With the formal scheduling system rendered useless, how could workers producing these items know where to place their priorities? Furthermore, this situation leaves no way to plan production to optimize batch sizes, minimize machine setups, or take other measures to improve efficiency.

■ A supplier is forced to respond to severe spikes in demand from its primary customer. Despite these seemingly unreasonable expectations, the supplier must maintain on-time delivery, high quality, and competitive prices. This is especially important now; with the industry rapidly downsizing, the supplier must continue to be seen as reliable in order to maintain preferred supplier status. Recently, this customer has begun to reduce inventories in response to pressures to lower operating costs. As a result, the spikes in demand have become much more pronounced; without the customer's buffer inventories, this supplier is now subject to the full range of variation of the customer's less-than-optimal operations. With operating margins driven to record lows, how can the supplier continue to contain costs as expected and remain profitable?

■ An assembly line for a complex, very expensive product is forced to regularly "travel" a great deal of its work, redirecting operations to install parts later in the production sequence when they are not available for installation at the planned workstation. Late supplier deliveries, receipt of deficient materials, or unscheduled maintenance has made it impossible to perform these operations as originally intended, forcing planners to delay them until conditions improve. The factory continues to move the product down the line, however, performing operations and installing equipment not dependent on the skipped jobs. This is not without consequence; traveled tasks are often completed in a suboptimal environment. With installation access much more limited because of the product's now advanced state of assembly, standard production tooling and assembly procedures can no longer be used. Time-consuming work-arounds must be applied, further disrupting operations.

These are not unusual occurrences; while their severity may vary, we found these same types of problems in a number of facilities across this industry. Again, we must ask why an industry that is so technologically advanced cannot fix these seemingly basic problems. Perhaps it is simply that those in charge have not yet been able to understand how best to approach them. In order to answer this question we sought to understand where this industry has been focusing its attention, and why it hasn't been successful.

Current Industry Focus

One of the areas that has historically received a great deal of attention across this industry is the measurement and control of direct labor, or "touch labor." We found labor performance to standard hours (a direct measure of the efficiency of touch labor within a factory) to be the most-tracked performance metric across this industry. We also found that this industry has waged an almost obsessive war on labor idle time in an attempt to influence this aspect of production costs. Facilities are developed, equipment is purchased, and labor agreements are reached in an effort to influence this, all apparently because of a belief that a factory whose workforce remains busy all of the time *must* be operating at the lowest possible cost.

In fact, we found this not to be the case.

After a detailed review of data across the different sectors of the aircraft industry, we found touch labor to consistently form only a minor overall part of the cost of production. In fact, with its entire contribution making up only 6% to 12% of the final product cost, we began to wonder if all of this attention is warranted. After all, with the considerable attention it has received over the past decades, any easy gains have likely been achieved, leaving little room for dramatic improvement. Furthermore, since touch labor is a relatively small overall contributor to production costs, any nominal improvement will not likely amount to a substantial degree of savings.[6]

Conversely, another area that this industry had begun to emphasize—the control of inventory levels within a factory—represents a large but relatively untapped cost driver. Because the cost of carrying these inventories adds substantially to the cost of the end product, a solid effort toward inventory reduction holds the potential to result in substantial cost reduction.[7]

Yet we found that, to optimize these savings, inventory *in all of its forms* must be attacked. This includes all unsold products—not only those items held in warehouses, but also those items in various states of assembly residing directly in the production shops. This much less emphasized form of inventory, known as work in process (or WIP), is very important, especially since it can sometimes make up the largest contribution to the costs of a production shop.

In order to understand the importance of WIP, we must first understand that a product continuously gains value as more labor, materials, and completed components are added as it progresses down the production line; for products associated with aerospace vehicles, this can be quite substantial (notionally depicted in Figure 1.2). Because of this increased value, the monthly costs incurred by carrying this as unsold inventory may accumulate to a substantial expense. We found that by the time a product reaches the latter stages of assembly, these carrying costs can make up *over 70% of the total cost of assembling the product* (as depicted in the figure, inventory makes up a higher percentage of the total cost of operations late in the production sequence).

This effect also serves to help explain the benefit of reducing manufacturing cycle times, or speeding the movement of

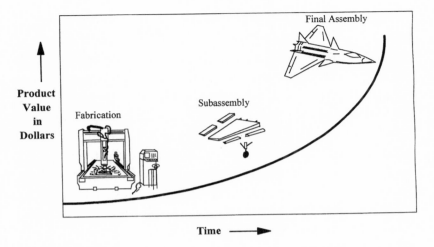

FIGURE 1.2 Increase in a product's value as it progresses down the production line.
Source: Based on a similar figure in *The Manufacturing Affordability Development Program*, Final Report.

a product through the production process. Any reduction to the time an unsold product spends at its highest state of value can have a major impact on overall production costs. Consider a $40 million aircraft that must be held at the flight line for three weeks to correct deficiencies or complete traveled work.[8] If we apply an annual carrying cost rate of 20%,[9] we can approximate a cost increase of almost half a million dollars *just for this one delay.* If similar delays throughout production can be eliminated, delivery schedules can be shortened to yield even greater savings. We can see that, even without the ability to fully quantify the spectrum of cost contributions, this reduction in carrying costs itself justifies attention to cycle time reduction.

To this point we have rationalized that inventory reduction and manufacturing cycle time reduction represent important cost drivers. Still, we found that a narrowly targeted attack on these areas may not be sufficient to drive substantial cost reduction. Despite a steadily growing emphasis on these areas, the anticipated savings have often not been realized. In fact, this emphasis may have led a number of facilities to encounter great difficulties. In one facility, a single-minded push to cut manufacturing schedules led to the stockpiling of those items that perpetually drove critical path schedules. While we did find that cycle times were reduced, this increase in inventories may have offset any savings, possibly even *increasing* overall costs. In another case, an emphasis on inventory reduction led to severe parts shortages across the facility, almost crippling production operations. From stories such as these, we began to see that something even more fundamental had to be addressed, something that these organizations had not yet begun to track.

Factory Variation: A New Focus

Further investigation of manufacturing cycle time data from factories experiencing these types of difficulties turned up a very revealing phenomenon. When we plotted the cycle times for each successive production unit (as depicted in Figure 1.3), we found there to be an inordinate degree of variation. While not clearly shown here (scales were eliminated to protect the confidentiality of the factory the data represents), the difference between the upper and lower points on this chart represents a value that is greater than 30% of the mean value. Subsequently, we found this same dramatic level of variation in cycle times at individual workstations across the production facilities.

This finding raises a number of important questions. With the degree of schedule uncertainty caused by such swings in unit-to-unit production time (as shown in the figure), how can a facility be successful in any attempt to slash inventories? In order to manage this schedule uncertainty, buffer inventories are vital to the factory's ability to support

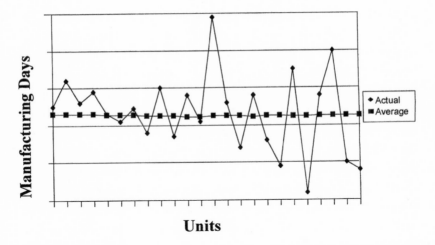

FIGURE 1.3 Variation in the amount of time required to assemble successive units.

Source: Based on a similar figure from *The Manufacturing Affordability Development Program*, Final Report.

production operations. Without the ability to forecast when parts and materials will be needed, Just-In-Time deliveries cannot be depended upon. On a more positive note, if the sources of this variation could be identified and controlled, couldn't manufacturing cycle times be dramatically reduced? After all, if one unit can be produced in the minimum time depicted here, it seems that they all could be produced in that same amount of time. As we will later show, it is only factory variation—most of which is avoidable—that prevents them all from being produced in this same amount of time.

Cycle time variation truly appears to be a new concept to this industry. While almost all of the facilities that we visited were collecting the component information, we found few to be tracking this data, and none using it as a metric to gauge their improvement. Many did not see this information as helpful, accepting the result simply as a part of the cost of doing business, much like the frequent need for traveled work and expedited materials. However, those that had been most successful in their improvement efforts quickly came to agree with us on the importance of this effect on any effort intended to drive down costs related to excess inventories and manufacturing cycle times.

From these and other observations, we began to recognize the limitations of efforts that seek cost reduction primarily by targeting cycle times and inventories. Only by attacking the more basic driver—production variation—does it appear that dramatic results can be consistently achieved.

Now consider the challenge of identifying the best approach for addressing this variation in its many forms. How can one determine which tools and practices are most appropriate, given the range of their effects when influenced by such diverse factors as product type, facilities, and production conditions? In what sequence should they be implemented? Like many others before us, we found that while one combination can result in a synergy that enhances improvement, another can lead to confusion and failure. As we observed the results of industry experimentation, we saw a definite pattern to successful change emerge, one that would help us wade through these complexities. This new understanding will be shown to form the foundation for the principle of variation management.

Uncovering Variation Management

A dramatic shift is seldom the result of one single event. Even in this industry so well known for its technological leaps, success is often the culmination of a series of lesser-known accomplishments. Because of their complex nature, it is only through this succession of successes and failures that a sufficient knowledge base can be gained to set the stage for advancement. Orville and Wilbur Wright's first flight at Kill Devil Hills, North Carolina, as well as Chuck Yeager's venture into supersonic flight are clear examples of this phenomenon. These feats, each forever changing the way we viewed manned flight, are widely seen as sudden changes. What is less commonly recognized is the importance of the gradual maturing of key technologies that served to make these events possible.[10]

Our observations led us to the conclusion that, in order to set the stage for a dramatic cost shift, the industry must again follow this path. With the immense knowledge base needed to evaluate the effects of even one tool on the many sources of variation, understanding the effects of a range of approaches would prove to be a daunting task. To complicate matters further, these tools and practices are rarely implemented separately; as a result, the complexity of understanding their individual contributions would require an immense database built largely through trial and error.

Fortunately, we came to realize that much of this experimentation had already been accomplished; we found that tremendous insight had developed across this industry through a range of largely independent successes and failures. Yet, since competition had precluded most from sharing these results, few could claim to have gained a clear picture of the broader lessons they offer. For years, this critical base of knowledge remained fragmented and disconnected, preventing even those that had been successful in demonstrating pockets of improvement from finding the path to broader results. How could these pieces of the puzzle be gathered together to provide the answers so critical to advancing to the next level of performance?

One of the key factors enabling this nation's revolutionary triumph over the sound barrier was the cooperative means by which the mission was accomplished. It was only through this partnership between government and industry personnel that key technical hurdles could so quickly be resolved. In fact, it was this aspect of the effort that was emphasized by President Harry Truman when he presented the 1948 Collier Trophy to Bell's Lawrence Bell, the Air Force's Chuck Yeager, and the National Advisory Committee for Aeronautics (NACA)'s John Stack.[11] With the need to quickly overcome such a large hurdle, the time for partnership is upon us again.

In fact, the principle revealed within this book could only have been discovered through just such an effort. We found that while a substantial base of knowledge resulting from over a decade of effort has developed from which to work, its fragmented nature has greatly diminished its benefit. Only through a unique third-party relationship offered by a government-led team could barriers be dropped to permit the necessary degree of access. With this team serving as the catalyst to bring it together, this same set of information became exceedingly more powerful. Comparison of the effects of different combinations of tools and implementation strategies from a large, diverse mix of companies across all three of the basic industry segments could now serve to demonstrate what has worked and what hasn't. In all, 17 separate facilities were chosen as the primary subjects, representing both the breadth and variety of products that are produced by this industry. This group included facilities within the following prestigious companies:[12]

Airframes/Aircraft Integrators

Boeing (Commercial Aircraft Company)
British Aerospace
Lockheed Martin
McDonnell Douglas

Aircraft Engines

GE Aircraft Engines
United Technologies—Pratt & Whitney

Avionics

Hughes Aircraft
Loral Corporation
Martin Marietta
Rockwell Collins Defense Electronics
Texas Instruments
Westinghouse

This book is intended to remove any doubt that widespread cost reduction can be rapidly achieved—even across such a diverse, complex industry as this—*if the principles outlined in this book are implemented.* Adding together these demonstrated pockets of excellence, much like assembling the individual pieces of a jigsaw puzzle, has enabled us to show the level of achievement that is possible. Through the lessons offered by both successful and less-than-successful attempts, we have been able to develop the complete picture needed for many to jump-start their quests for cost reduction. This understanding can be used by others to keep them from wasting time and resources pursuing initiatives that are not most appropriate for their situations.

The next chapter is intended to arm the reader with a historical perspective in order to enable a full understanding of why this principle has remained elusive for so many years. We will show that the current practices of this industry did not spring up overnight, that they instead are the result of almost a century of evolution within an environment that has experienced periodic causes for change. In fact, we will show that specific concerns that have been raised decades ago remain largely unaddressed today. We will also show that the solutions to these issues are not new, but their best course for implementation has not been widely understood.

CHAPTER 2

The Evolution of Manufacturing in the Aircraft Industry

The transformation of the aircraft manufacturing industry from its humble beginnings to its present stature has been nothing less than phenomenal. Early pioneers would undoubtedly marvel at the sight of the modern airplane, awed by both its tremendous capabilities as well as its growth to prominence in our society. They would likely be no less impressed by the technological advancements of the processes by which airplanes are produced. Automated machines have replaced lathes, drills, and saws to shape enormous parts to exceedingly fine tolerances.[1] Advanced composites facilities have displaced the old plywood fabrication shops, with automated tape placement and computerized autoclaves forming high-strength, complex structures with precision and repeatability. Electronic "pick and place" machines locate transistors and other components on circuit cards with dexterity and repeatability that can't be matched by human hands. Still, despite all of this technological advancement in tooling and equipment, it might also be noted that many other practices have remained surprisingly unchanged for almost half a century.[2]

The manufacturing of aircraft actually did not start with a classic production line at all—in fact, this innovation was not embraced until decades after the Wrights' historic flight.[3] Instead, much like the automotive industry, aircraft manufacturing has its roots in what has since been termed "craft production."[4] Under this concept, the production of airplanes was highly dependent on the skills of experienced artisans. Often using little more than rudimentary woodworking tools and only the simplest assembly fixtures, a small team built entire airframes from start to finish (see Figure 2.1).[5] Standardization was minimal; components were hand-fitted such that even similar items were unique. Without the limits imposed by standardized tooling and procedures, airplane designers often found themselves free to press the limits of existing technologies to meet the highly individualized specifications of their customers.

This approach of fashioning aircraft one at a time worked very well for a new invention that relatively few

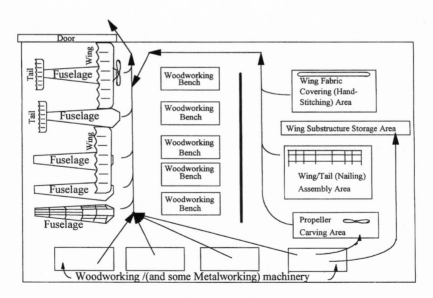

FIGURE 2.1 Typical aircraft manufacturing operation of the 1920s.

people had the money to purchase, or even the courage to climb aboard.[6] The early airplane was seen largely as a folly—a frail, high-risk contraption appealing only to a few daring individuals wishing to experience the thrill of "slipping the surly bond of earth."[7] With only 400 aircraft built by the entire U.S. industry in the year prior to the country's entry into World War I, the market was not sufficiently mature to support mass production.[8] Many were individually commissioned for highly specific functions, often with customers that expected this personalized attention—and that could well afford to pay for it.[9]

Over time, manufacturers became very effective at operating in this manner. Their strong reliance on craftsmen held down the cost of facilities and equipment, while permitting great production flexibility. While manufacturers could readily respond to the exacting demands of the customer, they were also able to maintain the type of quality workmanship that these products required.[10] Even with the broad range of products offered under this make-to-order approach, only the simplest of production management and oversight systems were needed to maintain order. In fact, it is for several of these same reasons that many have clung to aspects of this approach even today.

Only out of dire necessity did the industry begin to evolve. On the eve of World War I, this cottage industry was ill prepared to support the first massive increase in demand for its products. When the British and French entered the war, many of the early American aviation entrepreneurs saw a market forming and rushed to fill the growing need for these vehicles in Europe. They were largely unsuccessful in ramping up their production lines to meet the demand. They found that the production methods that had been so effective within a small niche-type market were much less capable within a high-rate manufacturing environment. Despite their best efforts, only a fraction of the requirement could be met, with orders instead filled from a larger overseas supply. As a result, even when the United States later entered the war, most U.S. pilots had to fly European aircraft. It became clear that an entirely different approach was needed.[11]

After the war, this need for change became more and more apparent as the marketplace for airplanes continued to expand. This growth was marked by two key influences: the increasing importance of airmail services offered by the United States Postal Service, and the rekindling of America's interest in aviation by the 1927 transatlantic flight of Charles Lindbergh. Serving to advertise the technological advances made since World War I, Lindbergh's flight reawakened America's interest in aviation.[12] Because of these demonstrations of aviation's potential for peaceful purposes, Wall Street began to take a closer look at what had traditionally been considered a risky, highly speculative industry, further spurring growth.[13]

This was also a time when aircraft makers began to move away from the use of wood and fabric in favor of more advanced materials for airframe construction. New lightweight aluminum alloys and fabrication processes permitted manufacturers to form their airframes into durable, complex shapes, thus broadly expanding the horizons of the aircraft designer.[14] It marked the beginning of an era that continues to this day—that of assembling aircraft by drilling and filling holes to attach aluminum skins to metallic substructures. With this shift, these vehicles came to resemble more and more closely their cousins, the automobiles. This similarity helped to make possible the cooperative efforts between the two industries that would be needed in the very near future.

<hr />

The advent of World War II set the stage for a number of key changes that would forever alter the way in which aircraft are manufactured. This event catapulted the U.S. aircraft industry to its current predominance, all in just over five short years. Prior to World War II—as late as 1939—the entire aircraft manufacturing industry was 44th in size among this nation's industries. By the war's peak in 1944, it had risen to become the world's largest manufacturing industry.[15] It is important to note that the automobile industry, along with several others, entered the business of aircraft production temporarily during the war, and accounted for some of this explosion. However, the regular aircraft industry produced about 90% of the airplanes used in the war.[16]

Most noteworthy among the changes that enabled this transformation was an accelerated shift from a dependence on skilled artisans—many of whom were now off to war—to a workforce composed of people who may have never before held a wrench. A new system took hold that could accommodate these workers, one that was also much more effective in meeting the very high production rates needed.[17] After the war, many of the skilled craftsmen returned home with the intention of going back to their old jobs—only to find that their broad skills were no longer the key to the production system that was now in place.

A New Era for Aircraft Production

While not usually associated with the aircraft industry, Henry Ford's philosophies have had a profound impact on the manner in which aircraft are produced. Ford's contribution came from his approach of breaking larger jobs into smaller, highly repeatable tasks.[18] Building on earlier gains in parts interchangeability and assembly line production, his successes in automobile manufacturing demonstrated the advantages of moving away from craft production. It was no longer necessary to assemble products from beginning to end as a complete unit; tasks could now be broken down to such a level that workers could repeat very narrow operations using specifically prescribed procedures. It was found that a worker assigned to relatively simple, repetitive jobs could become very efficient over time, yet still maintain or even improve the quality of the products. No longer constrained by the need to custom fit parts on assembly, the overall speed and capacity of the production line were dramatically increased.

During World War II, these innovations were broadly transplanted to the aircraft industry. Assembly and fabrication tasks were made to be more repetitive, with parts fabrication often scheduled in batches. Airframes that were once pieced together in single assembly jigs were now broken down into subassemblies and built up within separate parts of the factory. The decentralization of the assembly process resulted in a much less cramped work area; this dispersion greatly improved physical access to the airframe, allowing more workers to concurrently perform their operations. While requiring more total man-hours per day, this approach greatly reduced manufacturing cycle times, dramatically increasing production output.[19]

The BDV multiline production system exemplified this transformation. It was named after the companies that had first developed and implemented this system on the B-17 program—Boeing, Douglas, and Vega (a Lockheed company)—three of the aerospace giants that continued to dominate the industry until the end of the twentieth century. Rather than adopting a straight-line approach, which was characteristic of an auto plant, this new system used a concept that minimized movement of large pieces of aircraft structure. Under this concept, assembly operations were laid out into a series of concentric semicircles, with each arranged largely based on the size of the components, as depicted in Figure 2.2. Final assembly operations were located nearest the factory doors, which were consequently surrounded by the larger subassem-

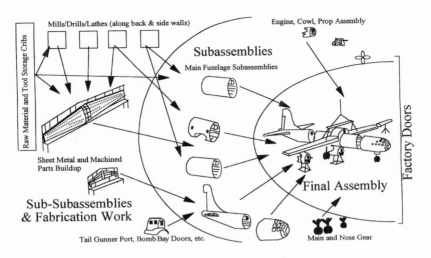

Mills/Drills/Lathes (along back & side walls)

Raw Material and Tool Storage Cribs

Engine, Cowl, Prop Assembly

Subassemblies

Main Fuselage Subassemblies

Factory Doors

Sheet Metal and Machined
Parts Buildup

Sub-Subassemblies
& Fabrication Work

Final Assembly

Tail Gunner Port, Bomb Bay Doors, etc.

Main and Nose Gear

FIGURE 2.2   The BDV Aircraft Assembly Line concept.

blies that fed into them. These were, in turn, surrounded by component fabrication areas or storerooms. This greatly reduced the factory travel distances for large aircraft components and assemblies as a function of their sizes.[20] This general approach continues to be used, even today.

> It was not practical to assemble bombers like cars and trucks on long, moving assembly lines that went past workers who attached more and more parts to the chassis until the final product drove out the factory door because bombers were too big and complicated. They didn't have the stable chassis of a car or truck. To build a bomber around its "chassis"—the main wing spar—would have been much too awkward. Aircraft workers couldn't stand on the floor and reach into whatever part of the bomber they were working on as it passed by, the way autoworkers could. To build a B-17 or a B-29 step by step on a moving assembly line would have meant a maze of ladders and scaffolds that moved down the line with the bombers as they took shape.
> —Jacob Vander Meulen, on the BDV production system[21]

Because extensive cross-pollination occurred between companies during the war years—touching almost every surviving aircraft manufacturing company—this assembly line system was largely adopted as the industry standard. Those across the industry benchmarked among themselves such functions as engineering planning and scheduling, resulting in organizations and production planning systems that were remarkably similar.[22] The fact that each evolved to the same end point is logical in that they were all faced with the same situation: The high sustained demand for these products mandated such an approach. It is interesting to note, however, that despite the fact that it had been developed in an environment that was much different from that seen since, this method of producing aircraft would become the model followed by most throughout the century.

The Complexities of Factory Management

This simplification of the workers' span of responsibility did not come without a price: By dramatically increasing the number of tasks that must be managed in an environment already stressed by increased production rates and product mixes, the job of managing the facility was greatly complicated. Rigid production control methods now had to be put in place to organize and coordinate long, variable sequences of operations. The result was a method of production that required a new mix of skills: a group of highly trained individuals to manage the overall production process, with a much larger group of more specialized workers to perform the manufacturing operations.[23]

The character of aircraft manufacturing organizations changed immensely to accommodate this new operational approach. A large number of departments sprang up to handle the various aspects of this now complex business. Coordination of work became a formal process largely accomplished through bureaucratic channels. No longer could an engineer make changes to a design by simply walking out to the production shop to discuss it with the chief builder, as Wilbur or Orville could with their mechanic, Charlie Taylor. Instead, separate teams of "liaison engineers" served as the interface between the product designers and their production shops.[24] With the creation of other departments to buffer workers from both suppliers and customers, rigorous communication channels became essential to enable the free flow of information.[25]

This organizational complexity led to a whole new set of problems, with production crises from the underestimating or overplanning of capacity, supplies, and raw materials becoming much more commonplace.[26] Yet, these problems were not unique to the aircraft industry. Even Henry Ford had experienced difficulty in managing the sprawling bureaucracy that was spawned by his automotive production line. His counterpart at General Motors, William Durant, had similar problems. It really wasn't until later when Alfred Sloan took over GM's management that today's means of managing mass production was established.[27]

In solving these problems, the aircraft industry suffered an additional complication in that it had not embraced these mass production approaches in their entirety. While the industry had adopted the basic tenets of mass production, it tailored the specifics to result in a hybrid approach. The multiline system represented one example where such changes were genuinely needed. In this case, the factory layout had to be adapted to suit the airplane's much greater complexity and size.[28] Other cases are not quite as easily rationalized, instead often lending the appearance that the industry simply could not bear to let go of the practices of its past. The most significant of these, a widespread continued reliance on craft manufacturing processes, has tremendously complicated the factory management process.

Some have argued that, because of the high-technology nature of the aircraft industry, craft production will never be entirely eliminated. As the customer continues to demand products that press performance to new limits, skill-based processes continue to be needed. While there is some merit to this argument, it is important to note that the processes requiring this level of expertise generally represent the exception rather than the rule. We have found that the majority of the processes necessary for producing even the most sophisticated aircraft are well within the state of the art. Airframe components are often produced to much the same tolerances as are used within the automotive industry, and it is consumer electronics—rather than military aircraft—that now set the pace for this manufacturing sector. Yet, we still find that customization is widespread across the industry.

In a typical facility, this customization can be seen from the buildup of subassemblies to final assembly and acceptance testing. It is not unusual for structural components to be individually hand-trimmed on assembly in order to make them fit into place. Electronic components often fail acceptance tests, requiring extensive troubleshooting and retesting. This type of special attention is not new; even in the case of the B-29, aircraft were routinely sent straight from the factory to "modification centers" where they received tens of thousands of hours of rework.[29] The added degree of uncertainty associated with these activities, in both schedule variation and material needs, clearly serves to increase the complexity of the already difficult job of production management.

This tendency to rely on the skills of individuals rather than on the regimented structure of true mass production extends even to the management of the plant. Despite the existence of rigorous management controls, the experience and instincts of individuals continue to be as important as the formal system under which they must operate. This led one early production manager to lament:

> The problem of scheduling aircraft fabrication is what one might call "complicated simplicity." It is not all academic or a series of numbers that can be multiplied and divided to get the answer. Much of the success of any scheduling system is due to the use of plain "horse sense" by the scheduling personnel.
>
> —Hector MacKinnon,
> production manager for Beech Aircraft, 1943[30]

As we will further discuss in subsequent chapters, it is this continued dependence on the specialized skills of workers, schedulers, and planners that has had the greatest impact on production efficiencies. The institutionalization of work-arounds has stifled the intended operation of formal systems to manage many of the greatest sources of variation. Even more importantly, this approach has spawned an ever-increasing acceptance of variation. As a result, it seems as if no sense of urgency has been placed on refining the primary methods of factory management.

The Proliferation of Work-Arounds

During the war, multiple plants—and even industries—frequently experienced simultaneous surges in demand, often leading to severe shortages of raw materials and components. With even one critical delay sometimes threatening to slow or even stop production operations, the aircraft industry was forced to take new measures to maintain some degree of stability. For this reason, many adopted the practice of prepositioning stockpiles of key materials.[31] Because of the immense cost, however, it was not possible to safeguard against all material shortfalls in this manner. As a result, the industry adopted other work-around procedures to minimize the impact of delayed receipt of materials.

One of these work-arounds was the establishment of a team of "follow-up men" (who later became known as material "expediters").[32] These people were assigned the task of identifying at-risk parts or materials and pushing them through the system to mitigate disruption to the awaiting assembly line. Because of the criticality of their mission, they were given great latitude in sidestepping established policies and procedures to achieve their goals. The result was impressive: Expediters were highly successful in minimizing the impact of scarce materials on production operations, thereby preventing major delays in delivering the aircraft to support the war effort.[33]

Following the war, many manufacturers continued their dependence on these approaches. Proven as a hedge against untrustworthy material and component suppliers during this time of crisis, they became widely accepted. Thus, these same practices were adapted for use during peacetime operations. Along with buffer inventories, expediters had become factories' insurance that production would continue to flow despite sometimes severe disruptions.[34]

We have found that these methods have not been as effective as many seem to believe. Despite the supposed insurance they offer, shortages continue to plague production operations. Even with hundreds of expediters and mountains of inventory, the single largest problem cited by many of these factories continues to be a chronic shortage of parts.

In fact, it is often *because* of these work-arounds that this problem continues to be so pronounced. In their haste, we found that expediters routinely circumvent formal purchasing and inventory control systems, leaving a substantial portion of material receipts and deliveries undocumented. Without *all* matériel transactions captured, inventory records are corrupted. The ability of these organizations to cost-effectively manage production operations without accurate inventory information was called into question even as early as 1944:

> Existing inventory records at most airframe companies are probably far from accurate and in some instances are recognized as being so inaccurate that they are seldom used for control purposes. Yet without accurate and timely inventory data no satisfactory materials control is possible. Without these data, also, total inventories must be kept large in order to avoid shortages in the plant.
> —1944 Harvard University study[35]

At the end of the war, this Harvard University study highlighted the cost impact that these types of behaviors could pose during the postwar drawdown. Others agreed with these findings, calling for substantial changes and cautioning that only through a much more rigorous approach to production could the industry efficiently make the transition to a peacetime environment.[36] Yet, despite the comprehensive findings of such a prestigious organization, many of these concerns seem to have gone unheeded for almost half a century.

Attempts to Control Manufacturing Inefficiencies

After the war, a new focus quickly drew attention away from these shop floor inefficiencies. A strong interest in new high-performance airplanes took hold. Jet engines, revolutionary new wing designs, and the subsequent leap to supersonic flight led to renewed demand for military aircraft and the creation of entirely new lines of passenger-carrying aircraft. Performance became paramount; cost was not a driving concern.[37] With this stimulation of long-term industry growth, along with some degree of industry consolidation, the pain that the aircraft manufacturers would feel as a result of a postwar contraction was not as severe as once thought.[38]

Throughout much of this century, this strategy continued to be highly successful. After all, it was this technological focus that offered an immediate competitive edge in both the commercial and military marketplaces. It allowed aerospace manufacturers to offer unique features that would differentiate their products from those of their competitors. Both commercial and military customers wanted the latest advances as quickly as possible and they were willing to pay for it.[39] Unfortunately, this focus led to the continued growth of manufacturing inefficiency. As one company executive observed in retrospect, "As business got better, efficiency got worse."[40]

This is not to say that manufacturing improvements were never attempted. There were cases where aerospace companies made valiant, successful efforts, sometimes in response to customer demands, other times simply in order to remain viable. Yet, these efforts tended to be aimed at achieving some preset goal; once attained, the efforts subsided. Many times these improvements lacked key provisions to prevent other problems from surfacing. As a result, many improvements were either short-lived or highly localized.

One published example of this phenomenon surrounded the Boeing Company in the late 1960s. The company was near bankruptcy and desperately needed to make improvements in order to slash operating costs. Production methods were far from optimal, and a downturn in demand for its products squeezed margins such that inefficiencies could no longer go on unaddressed. In response, the company pulled out all stops. After a detailed assessment, it took strong action to reduce its inventory stocks. Excess tooling and machinery were sold off, improved work practices were implemented, and tighter inspections were instituted. Furthermore, production flow through the factory was scrutinized and streamlined such that production schedules could be compressed.[41] Overall, these initiatives yielded sufficient improvement to save the company.[42]

Still, these actions did not completely eliminate waste. For instance, because suppliers were not made a direct part of the overall approach, their late deliveries led to part shortages that continued to plague the assembly line.[43] Despite a large degree of improvement, the use of work-arounds and inventory buffers continued to be important in permitting delivery schedules to be met.

It wasn't until decades later that a concerted effort to control these areas of manufacturing inefficiency began to spark the aircraft industry. Pressed to reduce operating expenses to accommodate shrinking customer budgets, manufacturers embraced a new round of tools and practices in the 1980s. Once again, lessons from the automotive industry served as the foundation for this movement. Yet, unlike what was seen in the past, these ideas did not originate at Ford or GM—this time they came from Toyota.

After World War II, the Toyota Motor Company began to experiment with its own concept of manufacturing, one that started with an understanding of the tenets of mass production, but adapted to work within the scope of its own environment. Toyota could not directly adopt the approaches used by the U.S. automotive giants if for no other reason than that there was a much lower demand for its products (at the time, this company produced far fewer automobiles in a decade than were shipped from Ford's Rouge plant in a single day). To compound the situation, the company had little access to investment capital, and had been recently pressured into an agreement with its trade unions that virtually guaranteed lifelong employment to the workforce. As a result, any solution would have to leverage the flexibility of this workforce to extract the highest degree of productivity from the company's limited facilities and equipment.[44]

Through trial and error, Eiji Toyoda and his production designer, Taiichi Ohno, proved out a number of innovative solutions to their problem. An emphasis on creating an environment of workforce flexibility and empowerment was central to their strategy; only with this could the company get the most from workers who would be tied to the company for many decades.[45] With the workforce trained for multiple disciplines, it would no longer be necessary for workers to remain idle between tasks. While waiting for a piece of production machinery, workers could now move to another station. Equipment was moved closer, quick setup fixtures were developed, and new procedures were put into place to facilitate workers' efforts. Toyoda and Ohno demonstrated that, by emphasizing workforce and equipment flexibility, the company could achieve astounding gains even without the efficiencies of scale of its competitors.[46]

This approach allowed Toyota to dramatically reduce the production cycle times and inventory levels needed to do business. With methods and procedures now in place to permit the fabrication of parts in very small batch sizes, work in process (WIP) inventories were reduced. Since Ohno focused on the elimination of activities that added no real value to the final product, a strong emphasis was placed on minimizing production rework. Using such methods as Statistical Process Control, processes were routinely monitored to keep them operating within the limits needed to maintain product quality. Continuous improvement, or *kaizen*, was adopted, where workers were asked to chase down the root cause of problems and identify means to improvement. Ultimately by driving down lot sizes and rework, buffers to schedules and inventories were dramatically reduced.[47]

Rather than focusing solely on the manufacturing efforts of its own plants, Toyota took additional measures to ensure the quality and timely delivery from its suppliers. As was noted earlier, external disruption would impact its operations just the same as if it had occurred within one of the company's own workstations. For this reason, Toyota implemented a system of coordinating the work of its suppliers. By emphasizing the development of a small, close-knit supplier base, Toyota could better ensure that they would not introduce significant variation into their own factories.[48]

Finally, a substantial effort was made in designing products for this type of manufacturing approach. With the reduction of machine setup time as a major consideration, for example, parts now were designed with common features that minimized the complexity of setup between different items. To improve quality, the processing capabilities of factory equipment now drove the complexity of product configurations. Cars were simplified, thus reducing the complexity of their assembly. In all, the vehicles were now designed to take full advantage of the lessons that had been learned in the factory.[49]

Through this approach, Toyota has been highly successful in competing on a price basis with the world automotive industry. In fact, its success in controlling cost while delivering vehicles with an unprecedented level of quality allowed it to attract a sizable portion of the world market. This ultimately created a crisis in Detroit; subsequently, U.S. automotive manufacturers were driven to adopt many of the same types of practices that had made their new competitor so successful.[50]

The success of these innovations has spawned their use in industries across the world. The tools and practices that had enabled Toyota to thrive despite seemingly insurmountable barriers are now widely seen as *the* secret to success. Faced with a strong customer demand to lower defect rates and improve product quality and reliability, many aircraft manufacturers set out to imitate these approaches. Yet, as was the case when Ford's approaches were first applied, this industry has not incorporated Toyota's approach in its entirety. Instead, much attention has been given to individual tools and practices rather than to the approach as a whole.

During the 1980s, methods such as Zero Defects, Quality Circles, and Total Quality Management targeted the improvement of product quality and factory waste.[51] Later, improved coordination during product development was emphasized; through a better design approach, a number of organizations hoped to drive down widespread disruptive factors such as engineering changes and defect rates. In the 1990s, many instituted programs to reduce their vast numbers of suppliers with the vision of closer working relationships and improved quality and responsiveness from those that remained. Other production streamlining innovations such as cellular manufacturing, one-piece production flow, and Just-in-Time manufacturing were attempted. Thus, in different combinations, this industry has attempted the entire range of tools and practices that were seen to contribute to Toyota's success.

This piecemeal implementation of the Toyota manufacturing approach—or "lean manufacturing"—has led to a much lower degree of improvement than many had hoped. Savings have rarely been broad-based; despite a limited number of notable exceptions, improvement has largely been isolated to individual workshops, or pockets of excellence within otherwise unaffected factories. As a result, many of those that produce aircraft and their components continue to be dogged by a complex web of manufacturing inefficiencies, and are generally forced to meet the rigorous quality standards of this industry largely as they have done in the past: through a strong dependence on costly, resource-intensive rework.

So why, then, has this industry continued to prosper? Even in the 1944 analysis that first pointed out these inefficiencies, it was impossible to foresee events such as the emergence of the Cold War and the tremendous growth of the commercial marketplace. Because of a major resurgence in government contracts, this industry did not face the true commercial marketplace pressures that had been feared. Many of the dire conditions that would have provided the impetus for change never materialized to the degree they were predicted.[52] Customer expectations, both military and commercial, continued to be based on superior performance rather than cost. Industry growth had largely been tied to its ability to continue its pace of technological advancements. As a result, there has never been a real need to change. Until now.

We are once again in an era where dramatic shifts in approach are necessary. Large cuts in the U.S. defense budget are impacting the aircraft industry at the same time that it must contend with a growing cost consciousness of its commercial customer. The customer now expects more than performance. Cost is now a factor of equal importance. Clearly, the time for change is upon us.

Current Industry Direction

Despite initial problems in implementing Ohno's innovations, many have not yet lost confidence in their suitability to the aerospace community. After all, modern-day aircraft manufacturers are driven by many of the same factors that first drove Ohno to develop his approach: Production volumes are much too low to justify the strict adoption of high-rate mass production techniques, and continued growth appears now to be tied to increased manufacturing efficiency. In the absence of anything better on the horizon, the industry does not yet appear willing to discard this very real potential for savings. As a result, most have continued to experiment with lean manufacturing tools and practices, yielding a variety of results.

The chapters that follow will demonstrate that there is, in fact, good reason to press forward with these innovative methods. Proceeding from a hard, scientific evaluation of this industry's lessons, we will show that problems in implementing these lean manufacturing tools and practices have most often been the result of underlying deficiencies in implementation. Attempting to spot-fix individual factory problems has meant that tools and practices are often not put in place as originally intended. Instead they are often used in a fragmented manner, leading to unpredictable outcomes. When applied within the context of their intended framework, however, these efforts have led to results that demonstrate the potential for rapid, dramatic savings. From this, we can conclude that they *are* suitable not only to the aircraft industry, but potentially to an even broader base of environments. It is clear, however, that a better understanding of this framework is needed to consistently yield these results.

As stated earlier, much has been written on specific Japanese tools and practices, but without a strong focus on the details of this broader context for their implementation. Benchmarking studies have largely targeted those cutting-edge steps demonstrated by leading companies. Much less attention is given to those steps that have only indirectly enabled improvement. Because of the incomplete picture that this has generated, the importance of these enabling areas can easily be underestimated. As a result, we have seen that they are frequently skipped, leading to a variety of difficulties.

In the next chapter, we will show that this industry's experimentation has yielded the data needed to complete the picture. The industry's successes and failures have pointed to a definite hierarchical order in which tools and practices must be applied to consistently yield a strong outcome. We will show that, by attacking the sources of manufacturing variation in a definite, organized manner, the magnitude and consistency of improvement can be greatly increased. If key areas that form the foundation for any factory improvement are first addressed, those cutting-edge solutions subsequently put in place will be much more likely to result in the dramatic degrees of savings anticipated.

CHAPTER 3

A Study in Manufacturing Affordability

The process of examining an industry's practices, even if only at a conceptual level, is by nature a major undertaking. For this reason, the type of study that we set out to conduct—one aimed at assessing the *details* of an industry's manufacturing methods—at first appeared almost impossible. Fortunately, we met with a unique series of events that ultimately enabled possibly the most revealing evaluation of the aircraft industry's current practices that has yet been undertaken.

We found that it is not only the complexity of the product, but also the nature of production that makes the study of aircraft manufacturing particularly a challenge. Upon entering an assembly plant, one is immediately met with an almost overwhelming scene. Vast spans of partially assembled, highly complex components scatter the building, with workers distributed in a manner that may seem even less deliberate. Even upon closer scrutiny, their activities can bewilder the untrained observer. Understanding the flow of operations can be a major challenge, with formal documentation for extensive sequences of operations often scattered and disconnected.

Without the right expertise, such a review could be time-consuming, disruptive, and possibly even futile.

The expansion of this type of effort across multiple sites further complicates matters. For instance, in order to begin to understand the complex linkages between specific improvement initiatives and their results, accurate and consistent data must be gathered. Yet, only through a rigorous data collection approach, one that independently validates methods and results devoid of potential biasing influences, is this possible. Reviewers need the freedom to roam about unsupervised, observing operations and questioning workers without attribution. Access to sensitive data is needed to validate and quantify results. Only through this type of thorough, rigorous process can detailed conclusions be reached. Yet, as many have found in the past, such an approach is generally not possible.

It is likely that this is precisely the reason why others have chosen to take an entirely different path; by limiting the depth of their studies, they have avoided many of these complexities. If they largely depend on those under study to provide "rolled-up" information, access to sensitive data is no longer necessary. The use of checklists and scorecards allows for further simplification; the need for subject matter specialists is virtually eliminated by presupposing potential findings and grouping them into broad, simplified categories. Thus, information becomes sufficiently basic to permit it to be gathered by generalists, or in some cases, even through the use of mailed questionnaires. Yet, this has shown to produce a much less sophisticated set of findings, and thus could not support the type of evaluation we sought to perform. To succeed in our goals would require the conduct of a much more rigorous review—and require us to gain an unprecedented degree of access to detailed information on both the successes and the failures across the industry. We knew that this was an unlikely proposition.

Yet this is precisely what we were able to do.

We found ourselves faced with a unique opportunity. The industry had become eager for solutions. With the Department of Defense equally interested, the stage was set for just the type of broad, collaborative effort that was needed to facilitate change. As a Navy initiative under the Department of Defense's Joint Strike Fighter program, we announced our intent to conduct a study to learn from the experiences of the companies within this industry. The industry responded enthusiastically, with many key organizations agreeing to permit precisely the type of access we needed. Because this study was to be led and controlled by an independent team, many of the traditional barriers to viewing sensitive and proprietary information were no longer issues. The time for cooperation was finally here.[1]

In fact, we found that this study couldn't have been better timed. Faced with new competitive forces, individual companies seemed to recognize a need to either rapidly reduce costs or face severe consequences. Racing toward this common goal, many had attempted a number of lean initiatives to reduce manufacturing costs. At the outset of this study we found that the industry had reached sufficient maturity in its experimentation with different mixes of a common pool of approaches that powerful examples of both successes and failures had been demonstrated. We could see that combining these individual lessons would place a wealth of knowledge within our reach.

In order to take the best advantage of this opportunity, we assembled a well-rounded team of subject matter specialists. Despite the broad exposure that we already had had to this industry, we recognized that our own experience base was largely as outsiders looking in. To balance the team's perspective, we enlisted support from individuals who could truly claim to understand the operations and constraints of this industry. Having worked as designers, production engineers, and factory supervisors, these individuals could truly claim to be insiders looking into the state of this industry. With this pool of experts from which to draw, we were ready to proceed.

For the purpose of brevity, we will not attempt to describe the backgrounds of each of these individuals. However, in order to give the reader a feel of the range and depth of experience resident within our team, we will review the background of two key members: Karl Stenberg and Don Garrity.

Karl Stenberg is widely respected for his expertise in the design and construction of cutting-edge airframes. Because of his long career within this industry, we found that he often could also offer a historical perspective on many of the phenomena that we encountered. Karl's experience spans many years, from the early postwar days when his role as a structural designer meant that he individually completed many of the activities that today are performed by large teams of engineers and planners. This included the design of a major airframe component (a wing), the structural analysis, and other activities leading to a complete production schedule for the product.

Karl's career continued as a designer in a range of positions with many of the aerospace leaders. As the head of composites research and development at McDonnell Douglas Aircraft, and later as the head airframe designer for the AV-8B Harrier jet, Karl captured his own piece of aviation history: He pioneered the use of advanced composite materials in modern-day aircraft with the successful development of the first all-composite production wing. In leading this effort, he became known for his skills in the design and certification of these cutting-edge structures, as well as his expertise in their manufacture and integration.

In contrast, Don Garrity's value to the team was in his shop-floor perspective of key manufacturing processes; more specifically, his expertise is in the fabrication of metal airframe components. Don can be regarded as a genuine expert; as an owner/operator of a machine shop supplying tooling and parts to major automotive and aerospace customers, his comprehensive knowledge of the methods and constraints of these operations was quickly recognized by those that we visited. This hands-on experience was bolstered by his background at Ford Motor Company, where he worked in manufacturing engineering, and from his government role in overseeing manufacturing activities of aircraft programs. With this comprehensive background, he brought to the team a balanced perspective of the complex issues facing this industry's assemblers and suppliers alike.

Don's observations were especially noteworthy in that he was able to successfully apply many of the lessons that we discuss throughout this book to his own shop's operations. As a result, his shop has become very successful, dramatically increasing customer orders. Because of the great satisfaction of one of these customers, Don found himself as the recipient of prestigious city and state business awards.[2] The demand for his products has so greatly increased that he has left his government position in favor of full-time pursuit of this much more lucrative venture.

Our access to similar talent in the areas of design and construction of electronics and jet engines, production and inventory control, tooling design and fabrication, supplier management, and a host of other disciplines was key to the success of our study. Because of this array of subject-matter expertise at our disposal, we could credibly assess the range of activities we will discuss in this chapter.

The stage was set for us to embark on the type of rigorous study of the aircraft manufacturing industry necessary to support rational, powerful conclusions.

Because of the immense amount of ground to be covered, we needed a means to quickly gain an indication of a facility's progress toward manufacturing cost reduction. We knew from our experiences with this industry that a number of activities at the end of the production sequence can provide this sort of insight. Traveled work operations, frequent expediting, and high levels of rework can point to troubles extending from a number of earlier operations or supplier deliveries. Even the absence of frantic work-around activity can tell a story; while it is sometimes an indication of a truly remarkable operation, it can also point to well-padded inventories and cycle times. Because of the value of these indications in quickly focusing our efforts, we began our reviews at the end of the line—with the assembly shops. We then proceeded through the factory in a sequence opposite to that of the production flow, coordinating our findings with teams dispatched to observe the operations at component fabricators and material suppliers.

To further prepare ourselves to gain the most from our onsite surveys, we asked for input from those we would review, hoping to leverage as much as possible from what the facilities had already learned about themselves. Armed with a much deeper understanding of their own operations, these organizations could make informed conclusions readily available to us. All that would be left for us to do would be to validate their data. Or so we thought.

Upon entering each facility, we were met by a team of plant managers, typically providing us with a two- to three-hour presentation on the accomplishments of their factory. Each team indicated that they had implemented a variety of new initiatives—almost always extolling dramatic results. Some related stories of quality improvements, their partnership with suppliers, and successes with Just-In-Time delivery techniques. Others emphasized their use of enhanced design tools and new manufacturing technologies that improved processing capabilities. Often we were presented with figures that detailed a substantial level of improvement against a variety of management metrics. Had our evaluation stopped there, we could easily have concluded that we had found the secret to widespread manufacturing savings, with no further need to continue our industry study.

However, our own observations often showed a much different picture.

All too frequently, we found that these inspiring stories represented a combination of optimism, misunderstanding, and oversimplification of results. Claims of success were more often anecdotal accounts than broader indications of success. Some had perceived that their own implementation of improvement initiatives was more rigorous than was actually the case. Claims of widespread savings were often unsupportable; they could sometimes be disputed using the facility's own data. In fact, in a few cases we were able to find data that could make the argument that factory costs actually *increased* as a result of their actions. Overall, we fell upon two disturbing trends: The actual results were generally less positive than we had initially understood, and, worse yet, these managers were sometimes unaware of the fact that their initiatives had missed the mark.

Our job became clear; we would have to separate fact from illusion and assess the true successes and failures resulting from this group of initiatives implemented in varying sequences and combinations. But how could we begin to gauge their relative degrees of success? What approach could we use that would prevent us from following the same path as so many who had fallen short on this undertaking?

Tracking Facility-Wide Improvement

From these initial findings, we recognized that a key part of our job was to develop metrics that could be used to measure the true impact of improvement efforts on factory performance. We found that only through the consistent application of these standards could we credibly assess the effectiveness of any of these initiatives. It was also important to learn from what we had already seen, steering away from one of the most common traps that had hindered the self-assessments of many of those that we visited: viewing results within too narrow of a context.

The best way to explain this trap is through the use of an example. Suppose that a new automated machine is purchased with the intention of both reducing manufacturing defects and increasing production capacity. When viewed within the bounds of its own workshop, this machine will almost definitely appear to greatly enhance productivity. By increasing production yields and decreasing the time and labor required for rework, it can be shown to have been a worthwhile investment. However, these savings may not be real; when viewed from a broader perspective, the benefits can sometimes be more than offset by hidden costs. For instance, this equipment may have been put in place at a point that does not represent a true factory bottleneck. If this is the case, it may offer an increase in throughput that goes beyond what is needed to support subsequent factory operations. The end result may be that it is largely left idle, with its usage insufficient to amortize its costs. Worse yet, the facility may feel the need to recover these costs by keeping the machine fully utilized despite a lack of demand for its output, resulting in the production of unneeded work in process (WIP) inventory. Ultimately these costs—each key contributors to the total cost impact on the factory—may not be visible to those determining the true savings associated with this equipment.

To avoid this vulnerability to sizable errors, we adopted metrics aimed at gauging *factory-wide* success. Also, since our study focused on those initiatives with the potential to quickly produce large savings, we steered away from metrics aimed at tracking lesser cost drivers, such as touch labor utilization. Instead, we came to view cycle time and inventory as key improvement measures (as discussed in Chapter 1, their minimization can substantially affect factory costs). We also sought to mitigate an important limitation of these metrics: Because their optimal levels are tied to such factors as factory size and product complexity, they cannot directly be used to compare dissimilar facilities. For this reason, we introduced another metric.

We found that the measurement of *manufacturing cycle time variation* to be critical to completing the picture painted by our metrics. Since it is strongly linked to the occurrence of factory disruption, it is a more direct measure of a facility's effectiveness in streamlining its operations. However, unlike product-based metrics of inventory and cycle time, this measurement can permit even seemingly dissimilar factories to be directly compared using such statistical relations as standard deviation. As a result, we viewed this as an excellent means of determining the effects of specific factory improvements on factory efficiency.

Common Enablers to Improvement

As we proceeded with our reviews across the industry, we compiled a list of tools and practices aimed at reducing manufacturing costs. As this list grew, it became cumbersome. Fortunately, we found that these initiatives readily fell into a series of six groups, or enablers, allowing us to track them based on their distinct areas of focus. We documented these as part of a structure by which we would conduct our reviews. This project structure is depicted in Figure 3.1. Using this breakdown as a baseline allowed us to refer to this large group of initiatives in an organized fashion. The structure is

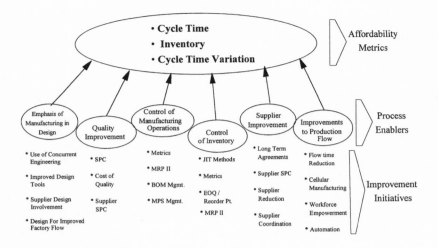

FIGURE 3.1 The project structure.
Source: The Manufacturing Affordability Development Program, Final Report.

capped off with our three metrics, cycle time, inventory, and cycle time variation, ensuring that all on the team applied the same yardstick to measure the facility-wide impact of each initiative. By using this as the baseline of our evaluations, we could ensure consistency in our approach as we moved from plant to plant.

As seen in the figure, there exists some degree of overlap across these enablers, with the same initiatives sometimes appearing in more than one category. The cross-link between Control of Inventory and Control of Manufacturing Operations is especially notable with two initiatives (MRP II and metrics) supporting the focus of both enablers. Despite this overlap, we believed it to be necessary to maintain these as two distinct categories because of the large, independent role of each in effecting manufacturing cost reduction.

We recognize that the titles of many of the initiatives within this project structure could lead one to a range of interpretations. In order to ensure that our subsequent discussions build from a common baseline, we will now pause to further clarify our intended meaning of the components of each of these enablers. Further discussions in Chapters 5 through 8 will provide additional clarification.

EMPHASIS OF MANUFACTURING IN DESIGN

The first enabler, Emphasis of Manufacturing in Design, is an area that has received a great deal of attention in recent years. Much has been written on the value of developing a more producible, simpler product that incorporates cheaper, more readily available standard parts. To better manage the number of initiatives in this category, we organized them into subcategories based on how they aim to achieve this goal. The first, *use of concurrent engineering*, represents a progressive philosophy aimed at increasing the involvement of personnel from all disciplines in the design process. Our focus is in its objective of eliminating the disruptive practice of throwing a completed design "over the wall" to the manufacturing organization, only to find that it can't be easily produced. *Supplier design involvement* is a closely related initiative. By allowing suppliers of major components to become involved in the design process, manufacturers hope largely to prevent production, delivery, and quality problems with their products. *Improved design tools* is composed of those tools aimed at enhancing product configuration in order to streamline fabrication or assembly of the product. Finally, *design for improved factory flow* takes the previous category one step further, focusing on that subset of design improvement tools that are intended to enable factory managers to streamline the way their factories flow. A further explanation of the distinction between these categories will be provided in Chapter 8.

QUALITY IMPROVEMENTS

Three types of initiatives have been broadly pursued within this industry for the purpose of measuring, tracking, or actively influencing the quality of components and products. The first, known as *cost of quality*, is typically mandated for operations producing products for the Department of Defense. Intended to provide the contractor as well as the customer with information concerning the costs associated with quality defects occurring during production, a formal database is dedicated to tracking the number and types of defects generated during production. Others, influenced by Japanese methods, have attempted an approach known as *Statistical Process Control (SPC)*, a statistical method of tracking, evaluating, and eliminating the causes of manufacturing process variation that can lead to product deficiencies. Finally, *supplier SPC*, or the use of SPC within the facilities of major suppliers, has been required by their customers as a means to measure and cause improvement to the quality of supplied parts.

CONTROL OF INVENTORY

Four different groups of initiatives were widely employed in an attempt to gain control of inventories. The first, *EOQ/Reorder Point*, represents a combination of two closely related inventory approaches. Economic Order Quantity (known as EOQ), is a method that identifies the most cost-effective ordering quantity based on a trade-off between volume price discounts and the cost of stocking and holding these quantities as inventory. Reorder Point is an approach used to maintain sufficient inventory stockage to prevent shortages between replenishments. Each of these is discussed in detail in Chapter 5. *Just-in-Time (JIT) methods*, on the other hand, represent a more contemporary inventory philosophy that encompasses a set of techniques that support the goal of having materials arrive at the factory just when they are needed for use, thereby eliminating the need to hold them in inventory. In order to determine organizations' degree of understanding of their progress in implementing these techniques, we sought to understand how they were tracking it. For this reason, we included *metrics* in this category. This is intended as a measure of whether such important areas as inventory and manufacturing cycle time are tracked, and whether they are balanced against each other to prevent trade-offs that could sacrifice total cost for improvement in one area. Finally, we included an automated production and inventory control system known as *MRP II* to gauge its effect on a facility's inventory control efforts.

CONTROL OF MANUFACTURING OPERATIONS

We tracked four different types of initiatives to determine their relative importance to the control of manufacturing operations. *Bill of materials (BOM) management* is intended to assess the degree to which a facility maintains an accurate understanding of what goes into the finished product, even as its configuration evolves due to ongoing design changes. We included an assessment of *master production schedule (MPS) management* for similar reasons: to gain insight into how rigorously those that we reviewed were maintaining an MPS that accurately reflects the sequence and scheduling of work flow throughout their facilities. The use of a formal production and inventory control system, *MRP II*, was again included here because of its role in managing factory operations. *Metrics* was included in this category for much the same reason it was included in Control of

Inventory: to understand whether facilities were ensuring that factory flow was not optimized through such measures as driving up inventories.

SUPPLIER IMPROVEMENTS

We sought to understand the contribution of four types of initiatives intended to improve the quality of parts and the timeliness of deliveries from suppliers. The effect of *long-term agreements*, or long-term contractual arrangements between supplier and customer, was explored. In conjunction with this practice, we sought to understand the effect of *supplier reduction*, or the practice of migrating to a much smaller, closer-knit supplier base. Because of the expectation that an improvement in quality can improve overall performance, we included *supplier SPC* in this category. *Supplier coordination* is a term that may be misleading; it is intended to identify the overall level of effort put forward by the customer to conduct business with a supplier. This includes both coordinating day-to-day activities with the supplier base and working with supplies toward long-term improvement in support.

IMPROVEMENTS TO PRODUCTION FLOW

This category includes a mix of initiatives aimed at streamlining production operations. *Flow time reduction* represents a series of tools and practices that are intended to reduce the portion of manufacturing cycle time associated with non-value-adding activities. This includes the development of fixtures and techniques targeting queue times and machine setup time, and also the rearrangement of activities to affect the time needed to transport parts across the facility. *Cellular manufacturing* is a method of optimizing the manufacturing flow within individual workshops by increasing the flexibility of the operation. *Workforce empowerment* aims to increase the involvement of workers to leverage their experience in identifying improved processing approaches. Finally, we looked closely at *automation* as a means to improve production flow. Because of the cutting-edge nature of this industry, we recognized that nearly all of those under study would be applying some level of automation; for this reason, we did not focus on typical levels of automation necessary for basic factory effectiveness. Instead, we sought to understand the effect of automation that extends well beyond the industry standard, in either complexity or breadth of application.

Surveying the Industry

As we began our reviews, we were quickly overwhelmed with masses of data. To make these data at all intelligible, we had to develop a simplified means by which they could be viewed. For this reason, we adopted a clear visual format. By condensing complex data sheets down to a series of simple symbols, we found that we could much more effectively scan for first-order indications of key interrelationships. Once identified, they could be used to simplify our process of analyzing data, providing us a quick visual indication of those specific areas shown to have high return potential. This table came to be known as our "comparability matrix."

Because of the importance we would place on this tool, we used a great deal of rigor in its development. We first constructed individual worksheets for each improvement initiative listed in the project structure, scoring them on a range of factors related to their relative effectiveness and degree of implementation at each facility we visited. For the purpose of simplicity, these scores were subsequently subdivided into three categories: weak, medium, and strong implementation.

When transferred to the comparability matrix, a weak implementation score was denoted by a blank, medium by a small
x, and strong implementation of an initiative denoted by a
large X (see Figure 3.2).

When we first began our review process, we believed
that each of the primary industry sectors—airframes, engines, and electronics—could not be directly compared. Our
impression was that the large differences in technologies
and manufacturing processes made them too diverse. We
also believed that, because of differences in their primary
customers—military versus commercial—they would have
been driven to achieve greatly disparate levels of cost performance. We were wrong.

Instead, we were astonished to find *no notable differences*
between these three sectors in making gains against our three
metrics. We found each sector to be experimenting with the
same types of initiatives—those outlined in our project structure. Regardless of whether we visited an airframe manufac-
turer, an engines plant, or an electrical board assembler, we
found that many experienced the same types of problems.
These initiatives being aimed at improving the efficiencies of
general factory processes, we came to understand that their
use as well as their outcome are not necessarily tied to the
type of product a factory produces. Thus, for the purposes of
our study, we discovered that their effects could be directly
compared across all three sectors.

Because of this, our analysis was greatly simplified. We
could now assess all of those we visited together within one
comparability matrix, providing us with a much larger pool of
data from which to search for trends. It is important, however,
not to lose sight of a remaining discriminating factor: While
the type of product is inconsequential, the *complexity* of a production operation is not. For this reason, we added a column
indicating facility complexity level, permitting this to be
taken into consideration during analysis.

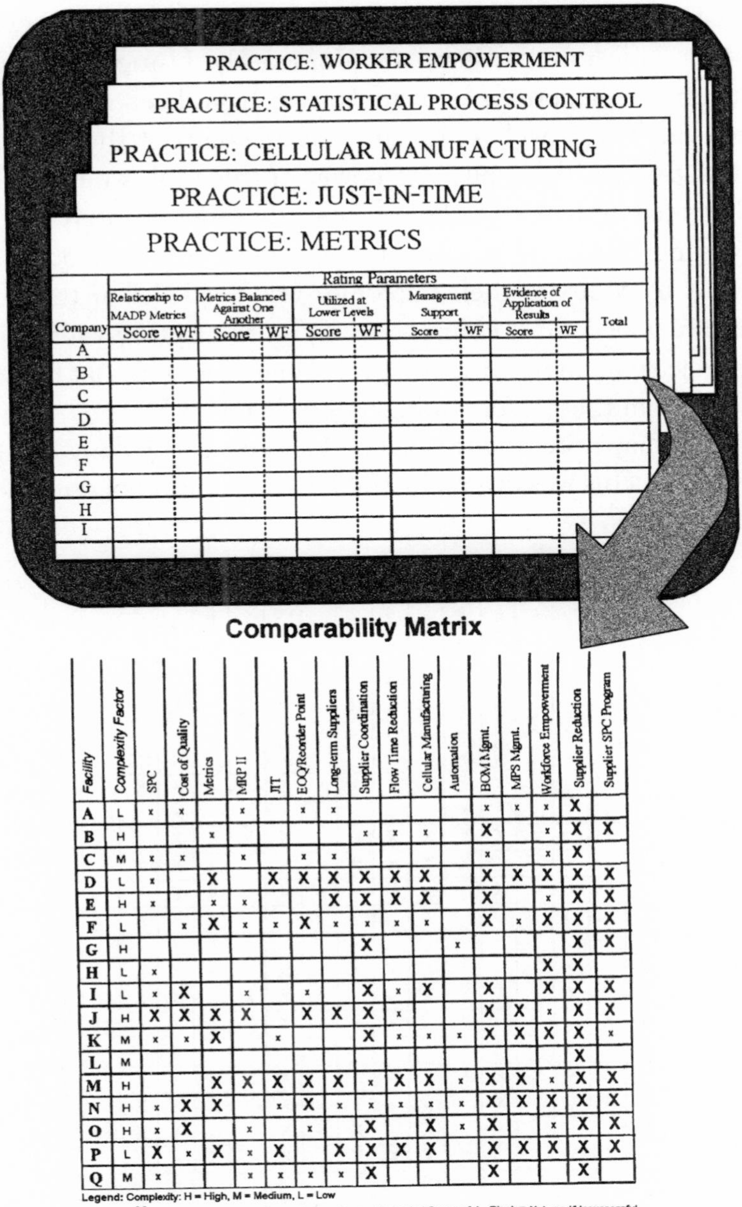

PRACTICE: WORKER EMPOWERMENT

PRACTICE: STATISTICAL PROCESS CONTROL

PRACTICE: CELLULAR MANUFACTURING

PRACTICE: JUST-IN-TIME

PRACTICE: METRICS

Company	Relationship to MADP Metrics		Metrics Balanced Against One Another		Utilized at Lower Levels		Management Support		Evidence of Application of Results		Total
	Score	WF	Score	WF	Score	WF	Score	WF	Score	WF	
A											
B											
C											
D											
E											
F											
G											
H											
I											

Rating Parameters

Comparability Matrix

Facility	Complexity Factor	SPC	Cost of Quality	Metrics	MRP II	JIT	EOQ/Reorder Point	Long-term Suppliers	Supplier Coordination	Flow Time Reduction	Cellular Manufacturing	Automation	BOM Mgmt.	MPS Mgmt.	Workforce Empowerment	Supplier Reduction	Supplier SPC Program
A	L	x	x		x			x	x					x	x	x	X	
B	H			x					x	x	x			X	X		x	X	X
C	M	x	x		x			x	x					x		x	X	
D	L	x		X		X	X	X	X	X	X			X	X	X	X	X
E	H	x		x	x			X	X	X	X			X		x	X	X
F	L		x	X	x	x	X	x	x	x				X	x	X	X	X
G	H								X			x					X	X
H	L	x														X	X
I	L	x	X		x			x		X	x	X		X		X	X	X
J	H	X	X	X	X		X	X	X	x			X	X	x	X	X
K	M	x	x	X		x				X	x	x	x	X	X	X	X	x
L	M																X	
M	H			X	X	X	X	X	x	X	X	x	X	X	x	X	X
N	H	x	X	X		x	X	x	x	x	x	x	X	X	X	X	X
O	H	x	X		x		x		X	X	x	X			x	X	X
P	L	X	x	X	x	X		X	X	X		X	X	X	X	X	X
Q	M	x			x	x	x	x	X			X				X	

Legend: Complexity: H = High, M = Medium, L = Low

Ranking: **X** = Disciplined Approach Across the Board, x = Somewhat Successful, Blank = Not used/Unsuccessful

Metrics: ↑ = Large Reductions, ↑ = Moderate Reductions, Blank = Small/No Reductions

FIGURE 3.2 Construction of the comparability matrix.

Source: Based on a similar figure from *The Manufacturing Affordability Development Program*, Final Report.

With this approach, we had developed the means to evaluate a wide range of data simultaneously. Using it, we can quickly see which types of initiatives have been pursued by each facility, as well as the relative strength of their implementation. Still, despite the power of this tool, we found an additional dimension to be necessary. In order to interpret the meaning of the differences in these facilities' approaches, a means to compare their outcomes was needed. For this reason, we added three columns to depict the factory-wide gains in cycle time, inventory, and cycle time variation that had resulted from these efforts (depicted in Figure 3.3).

You may note that this figure does not identify any facility by name. This was driven by the sensitive nature of our study.

Facility	Complexity Factor	SPC	Cost of Quality	Metrics	MRP II	JIT	EOQ/Reorder Point	Long-term Suppliers	Supplier Coordination	Flow Time Reduction	Cellular Manufacturing	Automation	BOM Mgmt.	MPS Mgmt.	Workforce Empowerment	Supplier Reduction	Supplier SPC Program	Cycle Time	Inventory (related to sales)	Cycle Time Variability
A	L	x	x		x		x	x					x	x	x	X				
B	H		x					x	x	x			X		x	X	X	↑	↑	
C	M	x	x		x		x	x							x	X				
D	L	x		X		X	X	X	X	X	X	X		X	X	X	X	↑	↑	↑
E	H	x		x	x			X	X	X	X		X			x	X	↑		
F	L		x	X	x		x	X	x	x	x	x	X		x	X	X	↑		
G	H							X				x				X	X			
H	L	x													X	X		↑		
I	L	x	X		x		x		X	x	X		X		X	X	X	↑		
J	H	X	X	X	X		X	X	X	x			X	X	x	X	X	↑	↑	↑
K	M	x	x	X		x		X	x	x	x	X	X	X	X	x		↑	↑	↑
L	M														X					
M	H		X	X	X	X	X	x	X	X	x	X	X	x	X	X		↑	↑	↑
N	H	x	X	X		x	X	x	x	x	x	X	X	X	X	X		↑	↑	
O	H	x	X		x		x	X		X	x	X		x	X	X				
P	L	X	x	X	x	X	X	X	X	X	X	X	X	X	X		↑	↑	↑	
Q	M	x		x	x	x	x	X				X		X			↑	↑		

Legend: Complexity: H = High, M = Medium, L = Low
Ranking: **X** = Disciplined Approach Across the Board, x = Somewhat Successful, Blank = Not used/Unsuccessful
Metrics: ↑ = Large Reductions, ↑ = Moderate Reductions, Blank = Small/No Reductions

FIGURE 3.3 The completed comparability matrix.
Source: The Manufacturing Affordability Development Program, Final Report.

From the start, we agreed not to link the identities of individual facilities that we visited to our detailed findings. It was only through this agreement that we could gain the type of openness needed to generate in-depth, validated results. For this reason, facilities are represented only by a randomly selected letter arranged in no meaningful sequence. While this approach effectively safeguards the sensitive data of those that we visited, we are satisfied that it provides the type of comparative information that can support our conclusions.

It is also important to note that one of our six enablers, Emphasis of Manufacturing in Design, has been left out of this matrix. Although it has the potential to provide a major contribution to manufacturing improvement, at the time of our study it had been utilized for only a relatively short time. Few of these development efforts available for our review had reached maturity; as a result, the improvements did not typically effect a facility-wide reduction in cycle time, inventory levels, or cycle time variation. Thus, this enabler could not be evaluated using our comparability matrix. Instead, another approach will be used later to evaluate its contribution.

A Potential for Dramatic Savings

From the comparability matrix, we can see that a number of facilities exhibit dramatic improvements against our metrics. But how great are these gains? We found that where there were sizable gains, they tended to be *very* large. The large arrows on this matrix represent a level of improvement greater than 25% because we found this to be a natural breaking point; improvement against our metrics was either well below or substantially above this mark. The maximum level of factory-wide improvement was astonishing—67% for cycle time, 80% for inventory, and 60% for cycle time variation.[3] Equally important, we found that this improvement usually occurred within a two-year period, with large gains often visible in the first six months. Clearly, something in these factories' approaches had led to dramatic improvement.

To complete this picture, we still had to find a means to show how these results affect manufacturing costs. In searching for an answer, we came across a critical dichotomy: While our metrics focus on the performance of a factory's *processes*, the measure of cost savings is generally tied to *products*. To remedy this situation, we devised a link between these two forms of data. This link is depicted as a comparison between the degree of cost savings we found for a facility's products and their performance against our three metrics, as shown in Figure 3.4.[4]

From this figure we can see a clear trend: that dramatic cost reduction results *only when all three metrics show a substantial degree of improvement*. This is important in that it strongly supports our conclusion that these metrics are direct indica-

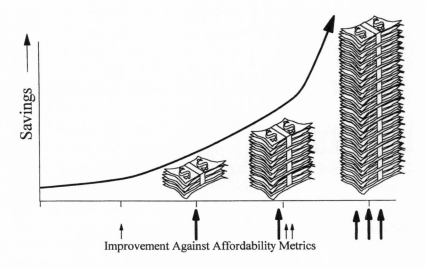

FIGURE 3.4 The relationship between the affordability metrics and savings achieved.

Source: Based on a similar figure from *The Manufacturing Affordability Development Program*, Final Report.

tors of substantial cost savings. Thus, we could reasonably argue that those facilities showing dramatic · improvement against our metrics are in the best position to control production costs.

Upon further scrutiny, we identified something even more revealing: that strong improvement in both inventories and cycle time is seen *only* when cycle time variation has been greatly mitigated. This supports our view of the importance of our cycle time variation metric, as well as the underlying premise of this book: *Only with the management of variation are dramatic cost savings seen.*

Turning our attention again to the comparability matrix, we can see a series of indications that point to the reasons for these savings. We see that those facilities with weak implementation of this group of initiatives—even for a large number of initiatives—have demonstrated little or no result. Only those exhibiting very strong implementation in breadth and commitment have realized the improvement against all three of our metrics, and thus have achieved major cost reduction.

With this understanding, we could now pare down our efforts. Since dramatic improvements in our metrics came about only through the use of strongly implemented initiatives, our focus fell on understanding how these had caused large gains in our metrics. We came to rely on our comparability matrix to identify how this industry's initiatives were applied to produce large improvements.

———————

In this manner, we assessed each of our study's six process enablers, evaluating the correlation between their implementation and the effects on our metrics. This analysis, detailed within the Appendix, points to a surprisingly well-defined methodology to success. It identifies a definite hierarchical order, one in which certain enablers must be first adopted to form the foundation for further improvement. It is through this successive approach to implementation that some factories have yielded large cost savings.

As depicted in Figure 3.5, the foundation of this framework is our enabler Control of Inventory, since only the facilities that had strongly and broadly implemented this group of tools have demonstrated large improvement against all three of our metrics. Control of Manufacturing Operations is next in line; dramatic improvement was again only seen when these tools had all been strongly implemented in conjunction with solid implementation of the previous group. The next level of enablers, Quality Improvements, Supplier Improvements, and Production Flow Improvements, all appear to show a real impact on cycle times, inventories, and cycle time variation, but only once the baseline elements previously identified are in place. The framework defined by these hierarchical relationships is illustrated in Figure 3.5.

The implications of this finding are astounding: No longer must a facility move blindly in its attempt to forge a path to cost reduction only to fail miserably. By focusing efforts where they can consistently result in success, time and resources are not wasted. For instance, from this framework it can be seen

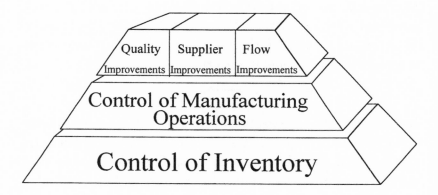

FIGURE 3.5 The hierarchy of process enablers.

Source: Based on a similar figure from *The Manufacturing Affordability Development Program*, Final Report.

that large investments in factory automation (a subset of the Production Flow Improvements enabler) are not likely to result in widespread success until control of inventory and manufacturing operations is achieved. Yet, once these foundation elements are solidly in place, this type of investment will likely demonstrate broad-based results. The same is the case for efforts in such areas as cellular manufacturing, supplier base reductions, and Statistical Process Control (subsets of Flow Improvements, Supplier Improvements, and Quality Improvements, respectively).

By using this model as a basis for determining where one's facility stands and which initiatives best suit its needs, the chances for producing rapid, dramatic results are greatly enhanced. As will be discussed in Chapter 9, these initial successes are key to creating the necessary environment for improvement. The increased enthusiasm they generate across the organization sets the stage for even stronger implementation of subsequent efforts.

The solid implementation of these building blocks appears to create the potential for even further benefit: We found that those facilities that had demonstrated considerable improvement against each of our three metrics had also developed a strong understanding of the need to include more progressive tools and practices in their product development efforts. As will be demonstrated in our analysis of Chapter 8, these facilities have shown a greater willingness to take on some of the more advanced design practices that promise to streamline the production flow of their next-generation products. This intuitively makes sense; the demonstration of success within current production operations should permit a facility to better understand the benefits that may be derived from this type of design focus. "In essence, these facilities are then better able to *pull* toward well-understood goals, rather than *push* toward some foggy destination".[5]

For this reason, the enabler Emphasis of Manufacturing in

Design was placed at the top of the building-block pyramid (Figure 3.6). This position is not intended as an indication that this group is either more or less important than the others. It is well documented that each of these design tools can have a dramatic impact on production operations. Rather, this is intended to depict that there is an added benefit to the design process by first implementing many of the lower-level enablers within existing production operations.

Dramatic Cost Savings Opportunities

So, how much cost savings can a facility expect to see if it applies this framework? Clearly, this question strikes to the core of any improvement effort; the larger the savings, the greater an organization's motivation for taking it on. We base our answer on a comparison of product costs as measured before and after these efforts had been applied.

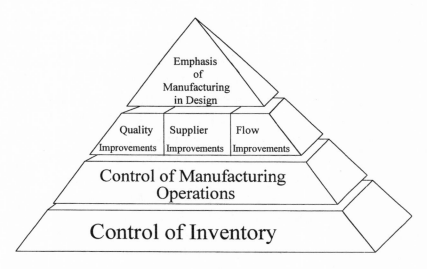

FIGURE 3.6 The framework for manufacturing affordability improvement.
Source: The Manufacturing Affordability Development Program, Final Report.

We first identified a group of items from each of the three different segments (airframe, engine, and avionics) whose cost reduction could be shown to be directly tied to the initiatives of our project structure. This limited the size of our set of data, since cases where a link to product cost savings could not be reliably validated were discarded. We then aggregated each of these demonstrated savings based on their relative effect if they could be assembled into a common aircraft. For this reason, our result must be taken as how much an aircraft could cost if each of this mix of components had been designed and built for a single end product. Thus, while no one product has yet seen this total savings, it is possible if this framework were to be fully adopted.

We were not at all prepared for the degree of savings that we found to be feasible. Familiar with the hard work that we had seen others perform to demonstrate only a modest cost savings, we originally speculated that the types

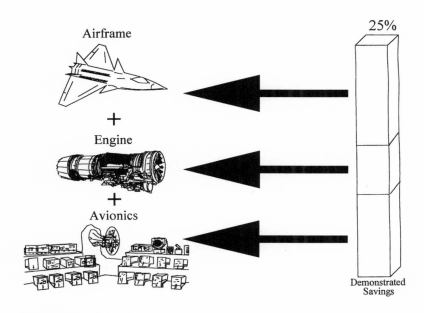

FIGURE 3.7 Total aircraft production cost savings.

of initiatives listed on our project structure would result in only a 4% to 5% savings in production costs. When we actually added up the savings potential (depicted in Figure 3.7) we found that *a cost savings of up to 25% was in fact likely.*[6] Furthermore, the true potential is much greater still, since we only drew from the limited set of data that we could reliably validate.

The intent of this chapter is to provide the reader with an understanding of how we conducted our study of this industry, and what made it especially remarkable. While we have provided an overview of our general findings, many more facts that serve to substantiate our overall framework as well as the best approaches to implementing specific enablers and initiatives were found. For simplicity, the details of what we discovered have not been described here; these have been left to subsequent chapters. We have, however, provided sufficient information to set the stage for a discussion of the principle of variation management.

CHAPTER 4

The Principle of Variation Management

As was described in the previous chapter, our study pointed to a specific framework that has repeatedly led to rapid, dramatic cost savings. At the core of this framework is a progressive attack on a key contributor to factory waste: variation. This variation manifests itself in a number of different forms, such as late deliveries, part shortages, underutilized equipment, and unpredictable production schedules. Its impact extends across all aspects of production—from the ordering of supplies to the assembly of the final products. We have shown that the success of this framework can be linked to the mitigation of this contributor, by successfully addressing key sources of variation, factory waste can be quickly and substantially reduced.

This forms the core of our *principle of variation management*. By adopting a systematic approach to progressively address the sources of variation that most drive factory inefficiency, companies can quickly realize dramatic gains. Yet, variation is a part of every process; it cannot be eliminated. There are far too many sources, with the influence of each on a given process essentially impossible to be completely controlled. Fortunately, we have seen that dramatic gains do not depend on the *elimination* of variation. Instead, dramatic improvement is possible by simply understanding the largest causes of variability and implementing a strategy to *manage* it.

To fully comprehend the importance of variation management, we must first gain a conceptual understanding of how variation affects a factory. To help us in this quest, let us consider a familiar situation:

Imagine yourself driving home from work. At first, traffic appears to be moving forward steadily, allowing you to travel at the speed limit. Suddenly, you see red taillights on the cars in front of you. The traffic slows to a crawl. You look around for an accident, but see nothing. There is no road construction; there are no entrance ramps. Nothing can be found that could have caused this delay. As you continue to creep along, you run the scenario through your head over and over again, trying to understand the cause. The radio traffic report indicates that there is no visible reason for the slowdown. This was not a usual bottleneck point, one in which the road narrows down, choking traffic flow. It takes 15 to 20 minutes traveling at this slow pace until the traffic abruptly speeds up again.

How many times have you run into this same situation, only to be puzzled and frustrated? You may be surprised to find that the cause of this problem is really quite simple. Picture an event in which one car slows down unexpectedly, perhaps as a result of an aggressive driver changing lanes too closely. A third driver following behind this event will likely be startled, perhaps hitting the brakes very hard in order to compensate. Since this driver must react based on very little information (the sudden appearance of taillights and a sense that the car ahead is slowing) that driver will likely overreact to ensure that there is no collision. Now consider all of the drivers behind this event, even in adjacent lanes; they are all likely to react in much the same manner, progressively slowing the traffic behind them. By the time you finally reach this scene, traffic is barely crawling forward, with the root cause— the aggressive driver—now long gone.

What can we learn from this scenario? Besides reinforcing one's understanding that aggressive driving can be counterproductive, we have also found that both the cause as and the end effects of this event are strikingly similar to what can be seen within a factory. It is evident that, as with cars on a highway, a certain degree of flow variation exists within a manufacturing operation. Cars each function independently, speeding up and slowing down to some degree even within a seemingly smooth traffic flow. The same phenomenon is inherent within any given production operation, with the time that it takes to complete a given task varying somewhat for each subsequent unit that is produced. When these events are linked together into a larger sequence, their individual variations accumulate in such a manner as to set the baseline pace for production flow.

Now suppose that something happens within the factory to disrupt one of these operations, such as a delay in the delivery of a key part needed for production, or an equipment breakdown. This isolated event can cause severe disruption to the output of an individual workshop. This will likely affect successive workstations, leading to late deliveries that could impede operations. The overall effect on this plant's production flow will be very similar to that which we observed on the highway: A ripple effect will likely result, with delays increasing in each subsequent operation. Much like the slowdown caused by the aggressive driver, a single event can result in an exaggerated effect. The result is a wild fluctuation in the time required to produce a product, leading to a constant scramble to try to make on-time deliveries.

In cases where this has occurred, a common symptom is

seen. Many of these facilities have developed a *culture of work-arounds*, or a series of informal approaches intended as stopgap measures to deal with immediate symptoms. By continuously rearranging production operations and expediting late parts, they ensure that promised delivery schedules are met. Even within the most disordered factories, production operations can be shifted to allow time to manufacture their products. Sufficient production capacity can be brought on-line to help deal with large fluctuations in schedules. Inventories can be raised to such levels to provide protection against stock-outs, despite a poor understanding of what supplies are really needed. This can permit deliveries to the end customer to be consistently made on time (depicted in Figure 4.1). Unfortunately, the degree of padding to production schedules, capacity, and inventories that is necessary to allow for this result can have a dramatic effect on the total cost of the finished product.

We found that these work-arounds are typically performed without concurrently updating the formal systems in-

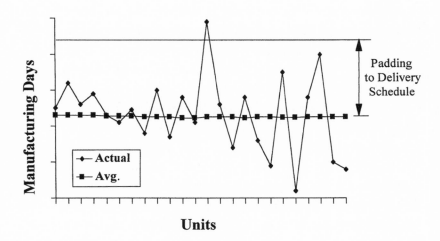

FIGURE 4.1 Cycle time variation as it relates to schedule padding.

Source: Based on a similar figure from *The Manufacturing Affordability Development Program*, Final Report.

tended to regulate the factory's inventories and production flow. In fact, it would be nearly impossible to track all of the work-around actions that we observed. Over time, the information on which these formal systems are based drifts farther and farther from what is actually occurring. As a result, the production plans and schedules the systems produce can become virtually useless. By default, the somewhat chaotic work-arounds that were originally utilized as supplementary measures now become the primary method of controlling production.

A large number of the facilities that we visited had fallen into this work-around trap. Without an understanding of the larger effects on their production costs, they seemed to have become satisfied with these results. Without total cost as a major concern, it's clear why they may have felt this way; despite the frantic pace of this work-around process, they have still been able to produce a high-quality product with deliveries generally meeting customer expectations. While this is, in fact, an impressive feat, it is now becoming apparent that it may no longer be good enough. The customer is now demanding more. Costs are too high. Production times are too long. We found that many are now beginning to realize that something must change.

A Framework for Controlling Variation

From this discussion, it is evident that the better factories are able to predict the events that will be faced down the road, the better they are able to prepare for and ultimately to respond to them. In fact, this simple premise forms the foundation of the principle of variation management. Without the ability to anticipate disruptions in traffic, one must react haphazardly when faced with them. If, however, a driver is able to anticipate the disruption in the traffic that lies some distance ahead, he or she can develop a clearer plan to compensate, perhaps slowing down much earlier. This will result in a more controlled response, eliminating the need to overcompensate. The progressive ripple effect throughout the system will thus be eliminated, allowing for flow stabilization and ultimately for full recovery. Even better, the driver who knows of an incident well in advance can choose to take a different route altogether. Similarly, within a production facility, the better the ability to forecast future stresses on individual manufacturing operations, the better the response can be controlled. For example, where past history indicates that inventory shortfalls are likely, a factory may plan to accumulate strategic buffer stocks as a solution. If a piece of critical equipment has a history of failure, increased preemptive maintenance actions may be in order. With a disciplined, formal system in place to understand and manage these sources of variation, the factory can avoid most traffic jams.

We discovered that the framework to improvement identified in Chapter 3 (depicted in Figure 4.2) provides a great deal of insight into how to go about cutting the waste associated with production variation. By successfully implementing the building blocks within this framework, key sources of variation can be progressively mitigated. From this we can see that certain sources of disruption *must* be dealt with before the full effects of improvements in other areas can be seen. Consider the highway analogy that was used earlier. Imagine that the traffic jam was found in the same location where a construction crew had shut down the entire left lane. Since this blockage forces traffic to merge into fewer lanes, a slowdown is imminent. In this case, regardless of any other sources of disruption, this most basic condition would still have caused delay.

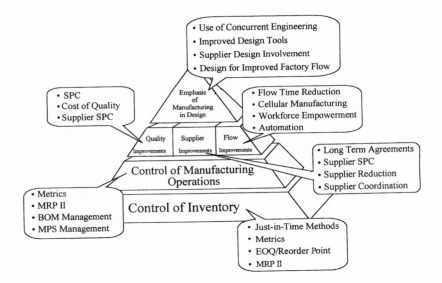

FIGURE 4.2 Framework for improvement.
Source: Based on a similar figure from *The Manufacturing Affordability Development Program*, Final Report.

It is exactly this effect that our framework to improvement illustrates. Within a factory, the control of inventory and production operations form the foundation of any approach to variation management. In essence, they are to production as the road is to automobile traffic; without first putting in place an adequate path, production flow cannot be optimized. The control of inventory is the most basic element, forming the foundation of this framework. Much like the narrowing of a road, the reduced availability of materials limits the pace of the overall flow. Without adequate supplies, operations can be seriously disrupted, leading to the widespread ripple effect described earlier.

We do not intend to imply that warehouses should be stocked full at any given time to prevent problems. As we saw during our reviews, large stockpiles of inventories do not necessarily prevent shortages of key items, and at the same time, these inventories bring with them substantial carrying costs.

Instead, inventory control is needed to ensure that the *right* materials are on hand when needed. This requires a disciplined program to keep track of both what is available as well as what is needed by the factory. With this, a facility can plan for potential roadblocks, balancing options for the most cost-effective solution.

Similarly, without first gaining a solid degree of control over manufacturing operations, how does one know what will be produced within any given workshop at any given time? Without this key information, it is difficult to plan to have the raw materials available to support these operations—as well as manpower and other necessary resources. For this reason, any approach to controlling inventories must work hand in hand with a facility's production control system.

With this foundation in place, a facility can now deal better with unexpected disruptions internal to its operations. Once the most basic roadblocks are removed, other approaches can now become more effective. By rigorously implementing production and inventory control systems (to be discussed in the next chapter), changes in demand can be "telegraphed" immediately across the line, keeping operations in lockstep. With an early warning of specific problems that will affect the production line, all of the shops across the facility can now adjust their plans to compensate well before a disruptive situation reaches them. Organized plans can be put into place, as opposed to the frantic expediting activity that would have been required without this notice. As a result, variation across the factory can be dramatically reduced.

Follow-on Enablers

While this improved degree of production and inventory control will alone result in substantial savings, the real value is in its enabling power for other follow-on activities. With the organization freed from the constant near-frantic expediting activities previously needed just to handle day-to-day operations, attention can now be placed on addressing some of the next-level drivers to variation. We found that once this basic degree of stabilization was attained, efforts to improve variation by focusing on such areas as quality, supplier issues, and specific production flow improvements became much more successful.

In principle, this makes sense. Consider the highway scenario once more:

> Imagine that marked police cruisers are dispatched to serve as pace cars during rush hour. Your first impression is that this approach is likely to slow down traffic, rather than to facilitate an improvement in traffic flow. In actuality, you find that while this presence does cause individual drivers to slow down, it also leads them to reduce disruptive behaviors which could cause ripple effects impacting traffic flow far down the highway. As a result, you see that it has led to a dramatically increased traffic throughput on the roadway.
>
> Now consider this same patrol car attempting to accomplish this at the point of a preexisting problem that degrades the overall highway conditions—such as the roadwork noted earlier. This more basic problem prevents the pace car from having any real effect on traffic flow. It just becomes stuck in the same traffic jam as the other cars on the road. Hence, you realize that the roadwork must be removed as an enabler to the effectiveness of the pace car.

This is analogous to the effects seen within a factory. Only when a basic degree of control is attained did we see that other improvement initiatives consistently lead to dramatic results. As an example, suppose that you are a shop foreman, and have managed to improve the production processes within your individual shop. Because of this, you have succeeded in eliminating a recurring delay in the delivery of your products to the next workstation in the production sequence. What is the overall effect to the factory?

Before answering this question, suppose that the same work-station to which you have improved your deliveries relies on a number of other shops in addition to your own. Now, suppose that these other shops have not delivered on time due to material shortages. Production will still have to stop and wait for these materials to arrive just the same as before you implemented your process improvement. From this it can be seen that, much like the patrol car stuck in a traffic jam, only with a solid degree of control over the more basic problems (in this case, inventory and production control) can your discrete improvement result in any real gains for the overall system.

This effect could be clearly seen in the application of the most popular of the flow improvement tools: the reduction in queue time (or wait time) between sequential work operations. In several cases, facilities had reduced miles of travel distance by simply moving workstations closer together. This not only produced the anticipated effect of reducing cycle times, but it was also seen to have had the effect of reducing travel time variation. With a simplified transportation process, there was less opportunity for disruption during the move. From a review of our comparability matrix (Figure 3.3), we found that even when substantial improvements like these were demonstrated, *they did not result in major factory-wide gains unless they were implemented in conjunction with the foundation elements of our framework.* As was seen in the case of the pace car, these types of efforts become most effective only once the first-order causes of disruption are cleared.

We observed similar effects in the implementation of quality control measures. As was previously noted, the most notable of these, Statistical Process Control (SPC), was seen to be strongly implemented within only two facilities. In both of these cases, a strong program was also in place for production and inventory control; *in this combination,* it was evident that the use of SPC had allowed for facility-wide gains in cycle times, inventories, and cycle time variation.

The use of SPC in understanding and controlling variation in manufacturing quality has been widely accepted over recent years. Its successes are difficult to dispute; this tool has been acknowledged as a key contributor to the production effectiveness for the Japanese automotive industry—a clear success story.[1] Unfortunately, there are also many examples of failure in implementing this tool. We found that a great deal of discipline is required to set up and maintain an effective SPC program—the same discipline that is essential for a strong system of production and inventory control. Perhaps this can help to explain why we consistently found SPC to be effectively implemented only where we also found strong basic production and inventory controls in place.

This observation appears to be consistent with the approach of the automotive industry. With the high degree of discipline reportedly found within these facilities, a strong degree of control over the foundation contributors of variation depicted in our model was very likely to have been in place at the time of SPC implementation. Conversely, SPC appears to have been largely seen by the aircraft industry (and likely by many others) as a stand-alone tool, with many expecting it to make rapid, dramatic improvements to their overall factory operations regardless of how it was implemented. Most seem to have attempted this tool without any thought given to the interrelationships identified within this book, probably contributing to the overwhelmingly less-than-successful results that we observed.

It must be noted that the cultures of these plants were key ingredients to their successes. While the type of culture that we found to be important is not widespread, it appears possible to develop it in most organizations. These observations will be discussed extensively in Chapter 9.

Control of Externally Driven Sources of Variation

Suppose that you are part of an organization that functions primarily as a supplier of components to another factory. You must optimize your production operations to satisfy the needs of these external operations, usually with very little insight or ability to forecast their future needs. Despite this lack of control, you must still operate efficiently, maintaining low-cost production in order to satisfy the price demands of the marketplace. You must constantly respond to changes in the customers' needs, often with little notice, resulting in a seemingly chaotic delivery schedule. Worse yet, if a major part of your business involves the production of spare parts, your required delivery schedules are even more difficult to forecast. With this in mind, must you resign yourself to providing poor response to the customer?

When armed with an understanding of the principle of variation management, the answer to this question is a resounding "no"! The basis of this principle is that dramatic improvements can be made through the use of internal controls *despite the existence of external constraints.* In this case, the suppliers must apply this principle in such a manner as to improve their ability to respond to the demands that will be placed on them in the future. We have seen that some have begun to do just this; by leveraging the information and resources that are within their control, they have developed a sound strategy to cost-effectively address this situation.

As we can see in Figure 4.3, the demand for an item can vary significantly from month to month, resulting in the types of peaks and valleys depicted. As a result, a supplier must maintain some sort of reserve capability—in the form of either production capacity or finished goods—in order to respond on time. We found that this is the manner in which

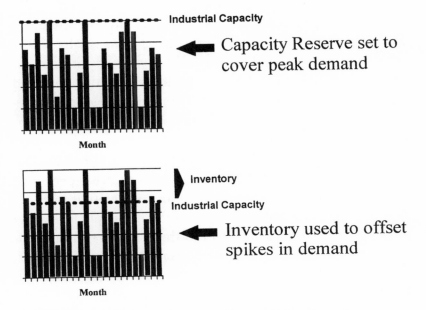

FIGURE 4.3 Typical responses to fluctuating demand: capacity versus inventory.
Source: Similar to a figure posted on the web site mantech.iitri.org/PUBS/DMC98 Pro/Index.shtml as of December 21, 1999.

many suppliers are forced to operate: with tremendous inefficiency driven by the variation of their customers that has simply become a core part of the business. Yet, we found that there is a better way.

As will be discussed extensively in Chapter 7, by better arranging the manner in which work is performed, manufacturers can dramatically reduce this variation. If a customer shifted its outsourcing approach to simply preplan the items to be produced by its vendors based on their common manufacturing processes, more of these items could be produced together on the same production lines. For groups of items where this degree of commonality is great (such as certain sheet metal parts, machined parts, etc.), process similarities can make them almost interchangeable in the production sequence. The variation in demand for an individual item would no longer matter;

with the production of these items now planned together, *the demand variation for the combined group of items* is what really matters. As depicted in Figure 4.4, the combined variation for a large group of items produced together can be demonstrated to flatten out considerably. Thus, demand can be better forecasted, production operations need not be frantic, and inventories and capacity can be reduced.

Later, we will show that a number of methods are available to take advantage of this methodology. While the most effective of these involve at least some degree of coordination with the customer, this is not absolutely essential. By simply leveraging those factors that fall solidly within the control of these facilities—such as how to arrange the factory and what types of orders to pursue—common items can be manufactured together, resulting in a dramatic drop in variation and overall cost.

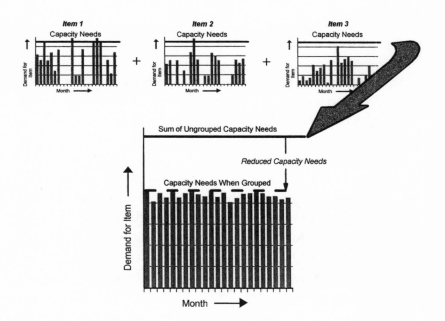

FIGURE 4.4 Aggregating demand to smooth spikes.

Source: Similar to a figure published on the web site www.dscr.dla.mil as of May 23, 1998.

This approach demonstrates the versatility of the principle of variation management. The same principle applies to all types of facilities; it just has to be viewed from a different perspective. While suppliers do not have the power to improve the causes of variation within their customers' plants, they are not powerless to address the problem. By leveraging their understanding of this principle, they can adapt their facilities by utilizing approaches such as the one just discussed. By doing this, they can continue to control their own destiny, even despite seemingly unreasonable customer demands.

Variation Measurement Techniques

Now that we have identified a general methodology to manage variation, how does one best track implementation of a specific solution? As with any problem, accurate measurement is important. To accomplish this, one of two basic approaches can be used: (1) measuring the end results of variation, or (2) direct measurement of the variation itself. In most cases, we saw that facilities were using the first approach, concentrating their measurements on factors such as labor hours and equipment utilization, and, in the more progressive organizations, on inventories. Much less frequently, we observed a comprehensive focus on measuring and tracking manufacturing cycle times. Only in a couple of facilities did we observe any real attempt to directly measure any factor more directly related to manufacturing variation.

The reason for this was fairly evident: The metrics that were most often used are fairly easy to link to some aspect of operating costs. Managers are already accustomed to dealing with equipment and labor costs. Even the costs associated with inventory reductions can be broken into quantifiable components that are easily understood (i.e., material costs and carrying costs, as discussed in Chapter 1). Variation is somewhat different; while readily measurable, it does not relate as directly to actual costs. As a result, on the surface it does not seem to be as important to the control of costs. Yet we found that its measurement does provide the manager with critical information about the facility's progress in cost reduction.

We found that, by routinely measuring and tracking a number of key metrics, a facility will better understand—and ultimately control—its own sources of variation. However, it is clear that each of these measurements must be continually compared against the others so that one area does not receive undue attention at the expense of others.

In our study, we found that a direct indicator of the degree of variation within a production facility is the measurement of cycle time variation. While improvements in inventory or cycle times were seen even where only minor cost savings resulted, it was only with a large improvement in cycle time variation that we found dramatic savings in costs. This stands to reason, since schedules and inventories can be reduced at the expense of other cost drivers. An improvement in cycle time variation, however, strikes to the core of addressing the underlying disruption; when the variation in production flow is flattened, the reason for padding these other areas is mitigated. From this, we can conclude that cycle time variation can be used independently as an indicator of dramatic improvement within a facility.

By measuring cycle time variation, a facility can gauge where it stands in its quest for improvement. As depicted in Figure 4.5, facilities that exhibited a small variation in cycle time (measured as a standard deviation in the range of 2.5 to 3) had exhibited a tremendous degree of improvement in the implementation of the first-order enablers to improvement (i.e., Control of Inventory and Control of Manufacturing Operations). This can be compared with standard deviations on the order of three to four times larger for those that had yet to master these enablers. As more and more of the initiatives

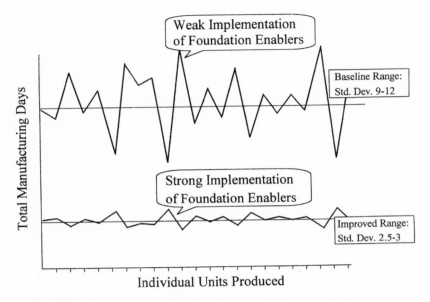

FIGURE 4.5 Effect of strong implementation of enablers on standard deviation.
Source: The Manufacturing Affordability Development Program, Final Report.

represented within this framework are incorporated, it is clearly possible to have an even greater effect on this metric.

With this in place, a number of other metrics can be used to further augment our ability to measure the gains in specific areas. As noted earlier, SPC is an excellent metric for tracking and managing variation in quality. The measurement of transportation distance provides insight into improvements in streamlining production flow, with shorter travel distances resulting in more deterministic travel times. Supplier SPC can be used to track variation in supplier quality. The measurement of manufacturing cycle times and inventories within individual production shops enables the direct measurement of these key cost drivers, providing information for additional analysis on specific cost impacts. Each of these can further hone our understanding of specific progress toward improvement.

When utilizing such a large group of metrics, caution must be taken to avoid the common trap of allowing the dominance of one metric at the expense of others. We visited a number of facilities that have demonstrated this shortfall. A common example is the strong emphasis of inventory reduction. As was noted in Chapter 1, some of those with whom we have spoken have come to believe that a policy eliminating *all* strategic inventories *must* be the immediate aim of any effective cost control effort. This emphasis is attractive because of the ease by which reductions can be equated to specific cost savings (also discussed in Chapter 1). Yet, from our discussions to this point, we can see that achieving real savings requires a broader emphasis.

We have seen a narrow focus in inventory reduction lead to widespread problems. In one case, a dramatic across-the-board reduction was made without first gaining strong control of manufacturing operations. This facility had not yet developed the capability to accurately predict which supplies were needed at any given time at individual manufacturing stations. The reduction of inventories eliminated the buffer that had previously enabled the system to function despite the lack of control. Once this safeguard was removed from the system, shortages of key items became commonplace. As a result, equipment utilization rates plummeted, cycle times skyrocketed, and cycle time variation increased dramatically. In short, this narrow focus almost brought the facility to a standstill—ultimately driving up the total cost of production.

We found that the means to avoid this potential trap had also been demonstrated: By tracking all important metrics simultaneously, a balanced focus is ensured. To support this conclusion, it can be seen from our comparability matrix (Figure 3.3) that all of the best-performing facilities had in place a strong system to track multiple competing performance metrics (captured in our project structure by the term *metrics*).

It is clear from these discussions that the control of variation within a manufacturing facility is paramount to achieving the rapid, dramatic degree of cost savings demonstrated to be possible in Chapter 3. It has also been shown that the framework to improvement revealed through the study of the aerospace industry provides a solid approach to attack the sources of this variation. Only by systematically addressing the sources of manufacturing variation can waste and inefficiency be controlled. Only by first putting in place these building blocks to form a solid foundation will a factory's overall improvement program be consistently successful.

Now armed with a greatly improved degree of control over production operations, managers can make business cases for major capital investments much more clearly. They will no longer be put in the position of having to make costly purchases of facilities or equipment just to achieve what could have been possible without them. Precious dollars can now be focused on areas where these investments are truly needed. This sharper focus can only help in further developing a company's competitive edge.

The next several chapters attempt to cover some of the specific lessons that we found contributed to the successful implementation of an improvement program. The reader must be cautioned, however, that even with this insight, this book is not intended to provide a detailed recipe for success. While we believe that we have identified the common denominators to a successful approach, success will still be dependent on the ability of managers whose job it is to adapt these to the specific circumstances of their own facilities.

CHAPTER 5

Production and Inventory Control: The Cornerstone Enablers

In Chapter 3, we demonstrated that the effective control of manufacturing operations and inventories is essential to mitigating key sources of production variation. Our discussions of the principle of variation management reinforced this observation; by offering a factory the means to anticipate changes down the road, individual shops can adjust their plans to compensate. No longer forced to constantly react to material shortages, late deliveries, and missed milestones, they can anticipate problems with enough advance warning to allow the development of well-staffed plans. By addressing potential issues in their earliest phases, these issues can be prevented from snowballing into larger, more disruptive problems. Thus, a great deal of unnecessary effort can be avoided. We found that this forms the basis to achieving rapid, dramatic cost savings.

All of the facilities that we visited had in place some sort of production and inventory control system. Why, then, did only a handful appear to be especially effective in mitigating the problems associated with production variation? We found that this is because the full potential of these systems is often not leveraged. Compromised by work-arounds, they often are relegated to tracking movements of inventory and work throughout the facility rather than used as a predictive means to driving down the sources of disruption.

Some have argued that this application is entirely sufficient, that it has long proven effective in ensuring that delivery schedules are met and manufacturing costs are generally in line with expectations. They argue that this has been accomplished while producing some of the most complex products in the world, meeting the most rigorous quality standards seen anywhere. What more should be expected?

In fact, much more *is* possible. Where these same systems have been used to facilitate better factory control, manufacturing costs have been slashed. Inventories prepositioned to protect against repeated unforecasted disruptions in manufacturing operations can be dramatically reduced. Production schedules no longer must be padded to compensate for wide variations in work completion. With variation dramatically reduced, the foundation is now in place for other enablers to address additional sources of variation.

We recognize that, with this discussion, we are touching on a very complex and controversial topic. With entire books dedicated to only a single aspect, we clearly did not intend for this single chapter to cover all that needs to be known about production and inventory control systems. Instead, we will focus on the role of controlling production and inventories in applying the principle of variation management. Within this context, we will explore the best practices that we found in implementing such a system. Ultimately, we hope that this will provide a sufficient basis for understanding which core steps must be considered, helping those embarking on improvement from stumbling as those before them.

Production and Inventory Control Systems

There exist a number of different philosophies concerning the most appropriate type of system to attain production and inventory control. Some argue that an MRP II style system is the way to go, with others holding an equally passionate view that a JIT-based system is the key to manufacturing efficiency. Still others swear by their own legacy systems, tried and proven within their own factories. We found strong evidence to suggest that each of these approaches has the potential to provide the high degree of production and inventory control necessary to enable rapid, dramatic savings. As a result, we could reach only one hard-and-fast conclusion: Regardless of the type, some sort of formal system *must* be used.

The real challenge is to select a system appropriate to one's own operating conditions and constraints, and to rigorously implement it. Our observations underscored this need to match the appropriate production and inventory control system to the complexity of work performed, demonstrating that such complex systems as MRP II are critical to complex operations, yet may represent an overkill for simpler factories. In fact, we have seen that the selection of too complex a system serves almost as an invitation to work-arounds.

While we do not intend to cover all of the types of production and inventory control systems, it is important to begin with a discussion of those that have been most successful within the plants that we visited. We will start our discussion with the much touted computer-based production and inventory control tools. We will then discuss methods shown useful in much less complex facilities, those producing items with simpler product configurations or production mixes. Finally, we will discuss some innovative hybrid systems that adopt features of both the simpler and the computer-based tools.

Our primary purpose here is simply to identify what seems to have worked well within this industry, as well as some of the real-world issues that we found in implementation.

COMPUTER-BASED PRODUCTION AND INVENTORY CONTROL

Manufacturing Resource Planning, more commonly referred to as MRP II, is a computer-based production and inventory control tool designed to automate key portions of the production management function within a factory. This is not to say that it can eliminate the job of the production manager; conversely, it is intended to strengthen that individual's ability to be effective in managing the facility. It is a push type of system, one intended to drive the pace of production flow by setting optimal plans and schedules to be followed by the production line. By doing so, it keeps sequential operations in lockstep; in terms of our earlier analogy, it develops alternate paths to keep traffic flowing smoothly despite the existence of hazardous conditions at various points on the road. Thus, it provides an overarching means of tying together activities across different phases of production, providing a tool to plan for contingencies and to minimize the need for expediting.

As a computer-based system, MRP II can analytically simulate factory flow, identifying the appropriate responses to pressures on the factory for each operation across the production line. As demand for finished products increases or decreases, this system can project the need for extra manpower, equipment, or material so that the production manager can develop plans to make them available. As can be seen from Figure 5.1, this information can be used to plan across short-term as well as longer-term time horizons. If the factory anticipates a delay of supplies from a vendor, the system can run what-if drills to determine the best short-term course of action to minimize disruption. The same information can also enable longer-term solutions, highlighting the need for changes in purchasing, manpower, capacity, or production scheduling. Master production schedules can be readjusted, maintenance plans shifted, or workers reorganized to deal with longer-term trends. When implemented as intended, MRP II provides a factory with a powerful tool to mitigate some of the most significant causes of factory disruption.

While there are a variety of MRP II systems available ranging widely in cost and complexity, each is based on a similar structure. As shown in Figure 5.1, these provide shop supervisors, planners, and managers across the factory

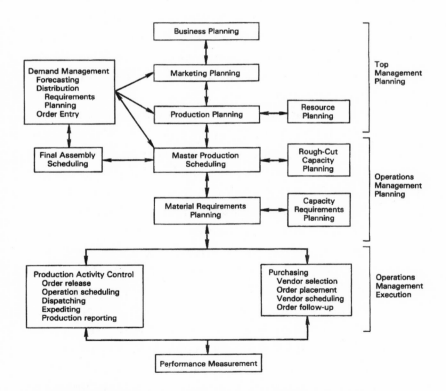

FIGURE 5.1 Manufacturing Resource Planning (MRP II) framework.

Source: Computer Based Production and Inventory Control by Smith, Spencer B., © 1989, Reprinted by permission of Prentice-Hall, Inc., Upper Saddle River, NJ.

with insight into a variety of activities. Individual modules perform the specific functions shown in the figure, yet work together to share common information applicable to the planning of facilities, materials, production shifts, supplier deliveries, and production schedules throughout the organization.

It must be emphasized that the success of an MRP II system cannot solely be attributed to the effectiveness of the hardware and software of the tool itself; equally important is how the system is implemented. Instead of treating it simply as a computerized tool, this system must be used as *the way of managing the business.* By managing operations through the system, the changes to inventories due to such causes as deliveries, expenditures during production, and discarded items are put into the system as they occur. Task completions and deviations from the master production schedule are continuously updated, providing an accurate picture of the factory's production status at any given time.

While data accuracy is important for all systems, it is especially critical for a computer-based approach. Human planners can make adjustments where mistakes are clearly made. Computer-based systems cannot make these adjustments; therefore, without accurate data, their output always becomes suspect. This will be further discussed later in this chapter.[1]

There is no shortage of examples of where MRP II systems have been put in place without apparent success. It is easy to see the evidence of where these systems have broken down: expediting, or working around the system to get parts where they are needed within the factory, is rampant. Compromised by these work-arounds, these systems can evolve into yet another mandated system requiring tremendous resources to operate while providing no real value to the company. When successful, however, MRP II systems have been linked to widespread savings.

LINE-OF-SIGHT SYSTEMS

While much has been written about complex tools for production and inventory control, they are not suited to all types of manufacturing operations. We found that less complex facilities, those with a low product variety mix and a fairly simple bill of materials (BOM), tend to thrive on much simpler systems. In these cases, manual systems appear to suffice, especially where the entire production line is within clear view of the production manager. It is for this reason that we refer to them as line-of-sight systems.

Under this construct, a part moves through the factory in such a short time frame that the tracking of its progress through each operation must be quick and straightforward. As a result, inventories are verified and production operations are tracked using simple, visual means such as charts and tables. Where most effective, inventories are maintained at very low levels. This allows any deviation from preset plans to become visible quickly, permitting it to be remedied long before it can cause serious problems to operations or delivery schedules. Because of this, there is no need to use computer databases to account for inventory levels or production status. In fact, a dependence on such a system here would likely slow down the decision process, and thus be counterproductive to the goals of the factory.

It is important to note that the latitude offered to the production manager by this approach should not be confused with laxness; despite its relative simplicity, a great deal of discipline must be present to prevent it from becoming compromised by work-arounds. As with an MRP II system, it remains essential to maintain very high data integrity, recording any deviations from the production plan. When unanticipated disruptions do occur, we found that the most successful managers have either acknowledged the need to make permanent changes or have taken strong action to get back on track as quickly as possible.

There are several features that characterize a successful application of this approach. The production work flow is sufficiently simple to permit visual systems to track the flow of parts through production operations. Individual operations are standard and simple, allowing any variation to become immediately evident. Inventory flow is straightforward, often with raw material delivered directly to the workstation for which it is needed. Completed items are not stockpiled; instead they are delivered to subsequent stations as they are produced. Machine setups are simple, standard, and foolproof so that part variation resulting from improper setup is minimized, as is cycle time from setups that take too long. In addition, batch sizes are consistent such that daily quantities can be easily scheduled.

While we sometimes found material deliveries to and from a shop operating under this type of system to be based on the Just-in-Time principle, this was not universally the case. Products could either be pulled through the facility when needed to meet the demands of subsequent operations, or pushed based on the preset flow of the rigidly regimented assembly line. As will be discussed further, success for this type of approach is largely dependent on its rigorous use; as is necessary for the MRP II system, it must become *the* way by which operations are managed.

HYBRID SYSTEMS

We also found a hybrid approach, one that combines the line-of-sight method just described with an overarching MRP II framework, to have been adopted by some facilities. When we first came across this type of approach, we were very skeptical; it did not fit at all into our own notion of how a production and inventory control system should operate. Yet, as we continued to scrutinize the gains made as a direct result of its implementation, we began to see it as a very effective means of managing a factory. In fact, we ultimately came to recognize that this approach can offer a sought-after balance: It provides the benefits of computer-based production and inventory control, yet without the degree of intrusion into detailed operations that this usually requires. This dramatically reduces the complexity (as well as the potential for workarounds) of a more traditional MRP II system.

At the fabrication shop level, this approach looks remarkably similar to the line-of-sight system. Operations are often set up such that they can be controlled through direct visual and verbal cues. These shops are left to operate autonomously, using a variety of approaches such as kitting as a mechanism to fill orders (to be discussed later in this chapter). What is different here is that this local system is effectively a component of a larger production and inventory system, with the orders that it fills driven by an overarching MRP II system governing the operations of the overall factory.

This construct permits both the design as well as the operations of the MRP II system to be much less complex. In allowing multiple line-of-sight systems to augment its functions at the local level (as depicted in Figure 5.2), the MRP II system can serve to pace the flow of operations between these shops. Data input is therefore much less frequent; one input is made once the complete set of parts to support the next level of buildup is produced rather than at the completion of each production step. By reducing the quantity of data and the frequency of its entry, the workload placed on the shop is minimized along with the likelihood of data entry errors. This observation helped us to understand the astounding success we found when MRP II was applied using this approach.

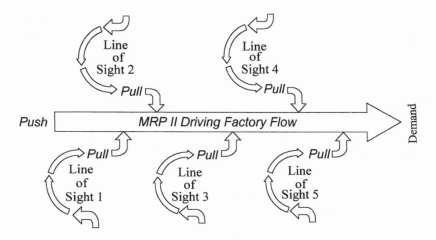

FIGURE 5.2 Hybrid production and inventory control system.

Factors Enabling System Effectiveness

Regardless of the type of system employed, we discovered dramatic differences in results based on how strongly the initiatives of these enablers were implemented. For those facilities displaying weak implementation, production forecasts were typically seen as unreliable. Very high inventories were insufficient to prevent frequent stock-outs, and the pace of production work-around activity could often be characterized as frantic. We saw an entirely different picture for those with stronger implementation. Stock-outs became a rare occurrence—even with inventory levels dramatically reduced. Individual workstations were no longer subjected to the degree of unanticipated surges that had previously disrupted operations; as a result, the receipt of raw materials and the delivery of completed items much more closely followed production plans. Padding to cycle times, production schedules, and inventory stocks were greatly reduced. Most importantly, however, the need for armies of expeditors was eliminated, along with the frantic pace of activity necessary to accommodate their work-arounds. All in all, the strong implementation of production and inventory control initiatives led to a much more consistent production sequence, and to a much more streamlined production flow.

After evaluating a number of different characteristics, we concluded that production and inventory control systems have led to the strongest results where we observed the presence of each of the following:

- A clear, consistent inventory control policy is in place.
- A baseline from which to build has been set.
- Key data is accurately controlled and maintained.

Facilities that had demonstrated the greatest degree of improvement against our three metrics had demonstrated strength in each of these areas. Conversely, those less successful were missing at least one of these characteristics. As a result, we concluded that each of these represents an area that should be strongly emphasized to maximize the likelihood of success.

ESTABLISHING A CLEAR, CONSISTENT INVENTORY POLICY

Our analysis of the comparability matrix of Chapter 3 revealed that there is a need to fully embrace *some* consistent policy of purchasing and holding inventory. We found, however, that the degree to which this is actually done varies substantially. While some facilities have adopted comprehensive inventory practices based on such progressive approaches as Just-in-Time, others had just begun to consider the use any formal inventory policy. We found that even large, prestigious organizations had sometimes only just discovered the method of Economic Order Quantity (EOQ), a concept that has been widely used for decades. Yet, most appeared to understand the importance of this step in setting the stage for continued improvement.

As depicted in Figure 5.3,[2] EOQ seeks the lowest-cost solution in ordering factory materials. It is based on the realization that multiple factors must be considered when striving for improved efficiency; more specifically, gains made in purchasing efficiencies must be balanced with increased inventory costs. In order to take advantage of economies of scale, a decision may be made to purchase more than the quantities necessary to support immediate factory operations. In other words, the total cost per item will normally be lower when buying in bulk. Under EOQ, however, this savings must be balanced with the offsetting costs associated with holding

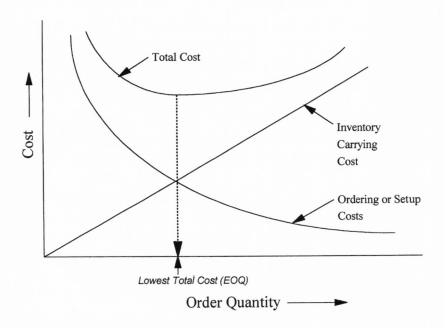

FIGURE 5.3 Economic Order Quantity curve in its simplist form.

these higher quantities in inventory. The figure depicts this trade-off with a series of curves. As shown, the point where these curves combine to yield the lowest-cost solution represents the Economic Order Quantity.

A further step is often taken to identify when these orders should be made. Reordering triggers can be established based on a factory's ordering approach. Orders can be triggered at predetermined time-based intervals, or when inventories reach preset stock levels. This latter approach is depicted in Figure 5.4.[3] As seen here, when inventories reach a level known as the Reorder Point, a new order for materials is made. This Reorder Point occurs at such a level that supplier deliveries will normally be received before safety stock levels are depleted. As seen in the figure, this safety stock is used to buffer delayed vendor deliveries and increased rates of consumption. While some may frown on this type of system because of its inventory component, we found that these costs can be far outweighed by the benefits of preventing matériel stock-outs—a condition that would otherwise disrupt assembly lines.

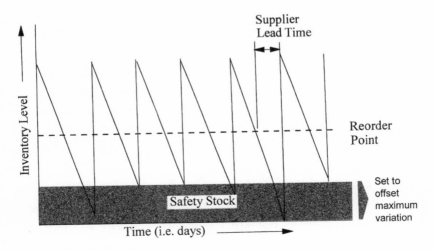

FIGURE 5.4 Reorder Point concept.

While the absence of a solid inventory policy limits a facility's ability to make cost-effective ordering decisions, we found that it can lead to an even more severe shortcoming: a poor awareness of factory inventory levels. A solid knowledge of inventories is paramount; it forms the basis for all other efforts that follow. We found that only with a detailed understanding of *all* inventories resident within the factory (both warehoused and work in process) can a facility find itself in a position to cost-effectively prevent part shortages before they precipitate work stoppages. Without this visibility, the efficient tracking and movement of materials to keep individual production shops moving efficiently becomes difficult. The alternatives are to extend production schedules or to buffer inventories—each of which can dramatically affect the cost of the product.

Many of those that we visited acknowledged this need, but still proceeded with such initiatives as cellular manufacturing, advanced factory automation, or MRP II systems without first demonstrating anything near the requisite inventory accuracy levels. Often, this was not intentional; we visited more than one facility that had developed a false sense that it had achieved a very high degree of inventory control. Some fell into this trap by assessing inventory accuracies based on *average* findings (where overstockages of some items are used to negate understockages of other items), thus painting a net positive picture where the true story was, in fact, quite different. This approach to accounting had concealed the fact that quantities of individual items were nowhere near their required levels. As a result, these facilities proceeded with the false belief that they had satisfied this important enabler.

In other cases, we found that this step was simply ignored. Instead of improving inventory control, extensive parts expediting systems were put into place to support production needs. We found that some of these have grown to an immense scale, with hundreds of individuals employed for no other purpose than to work around normal channels to keep production operations moving. It is not difficult to understand why these organizations consistently display a great degree of cycle time variation—and have been consistently disappointed in their cost reduction efforts.

A variety of methods have been successful in attaining high inventory accuracies, the most common of which is the exertion of direct physical control. Based on the apparent philosophy that, given the chance, the workforce will attempt to circumvent the system, this approach requires all inventory to be locked up in central cages. Consumable items, such as fasteners, drill bits, and parts kits, are strictly rationed out based on proven need. In some cases, even WIP is handled in this manner. Here, as a component completes a major operation, it must be checked into a central control crib to be reissued only when specifically needed to complete a subsequent operation.

We did find facilities that achieved a similar degree of inventory accuracy with much less physical control, instead focusing on changing the mind-set of the workforce. Implementing changes to assure that materials will be available when needed, they attacked the core reason for maintaining hidden stockpiles: By instilling a firm belief in the inventory system, the need for hoarding items was eliminated. With the demand for items now much more stable and predictable, a variety of other more advanced material management practices such as kanban could now be put in place. This approach to changing workforce culture will be discussed extensively in Chapter 9.

Occasionally, we found that these facilities employed a pull type of inventory system that permits vendors to directly replenish customers' production shop bins with shop consumable items (such as fasteners, drill bits, and machine cutters). Under this approach, the vendor is made responsible for closely tracking the shop's usage of each of these items, matching a delivery schedule to satisfy this demand. This streamlines the replenishment process in two ways: It eliminates the factory's costs associated with holding and managing these inventories, and it furnishes the vendor with firsthand usage information from which it can streamline its own process flow. Those facilities that use this concept consistently demonstrated very high parts availability in conjunction with improved costs.

For more complex production environments, we found that the practice of kitting components allows workers to efficiently provide all of what is needed to subsequent workstations. With this approach, a crate is provided containing compartments to hold a complete set of parts—those items needed to support the production needs of the next end item. Only after each compartment is filled is the kit sent forward. This approach allows workers to readily keep track of what is left to produce, enabling them to prioritize their operations without the need for elaborate planning activities. Thus, workers can play a key role in maintaining low levels of WIP while demonstrating high inventory accuracies, and without the inefficiencies associated with strict physical control.

ESTABLISHING A SOLID BASELINE

The cornerstone of any effective production and inventory control system is the establishment of a stable baseline from which to build. A detailed understanding of inventories forms a major part of this effort; by first identifying the items that exist within a production facility, the foundation is laid to further track them as additional materials are purchased or other items are manufactured. The same baseline is also needed for the other key activities conducted within the facility. In order to identify what must be purchased, the specific needs of each manufacturing operation must first be defined. The same applies to manufactured items; before setting up detailed production plans, the requirements for specific components from preceding manufacturing shops must be known.

While these observations may appear to be intuitively evident, we have found that many of the elements important to establishing this baseline are not routinely in place. It was common to find such core tools as the bill of materials (a map of the components forming the final product) to be inaccurate. To complicate matters further, many of these same facilities had not first established an accurate count of the components that already reside within their production shops. Without reliable inventory and configuration information, how can their management systems credibly prioritize the parts that must be purchased or manufactured to support the production lines? Think of the confusion to which this can lead: A system is put in place to manage production operations without the requisite insight into what goes into each product or what is on hand from which to fill orders. When viewed in this manner, it becomes much clearer why we found that these systems inevitably fail to operate as intended.

Those that were most successful made a substantial effort to ensure the rigor of this information base. By auditing their inventories, bills of materials, and manufacturing steps prior to proceeding with plans to improve their formal production and inventory control systems, they consistently met with success.

It was very common to see facilities activating a complex computerized production and inventory control tool without first establishing the baseline from which it could begin. The comparability matrix of Chapter 3 gives us a strong indication of how prevalent this practice is within the aircraft industry: Of the 10 instances where MRP II was attempted, only four cases had both the bill of materials and the inventory control activities we measured been first considered. In the cases in which this control had not been attained, the MRP II system was expected to maintain an accurate picture of inventory levels and work sequencing across all stages of production—despite the fact that it had not been provided a correct point from which to begin. We found that this strategy always led to an ineffective system.

CONTROLLING KEY DATA

Once a baseline understanding of production and inventories is established, its integrity must be maintained over time. Yet, in order to achieve this result, the system that is expected to maintain this information must be utilized in performing day-to-day operations. Where it is not an integral part of everyday business, its effectiveness can quickly wane. If it is viewed only as an administrative tool, care will not be taken to maintain its accuracy. As a result, unrecorded work-arounds will quickly cause its representation of production operations to diverge from reality. As its output becomes less and less useful, the workforce will begin to ignore it altogether.

We found that this type of failure can come from good intentions. Once the need to set up a new production and inventory control system is identified, a certain mood of urgency often prevails. Eager to adopt the system, facilities may turn it on before some key enabling activities have had the chance to sufficiently mature. Until the time is taken to set a solid baseline from which to move forward, the system is not yet capable of managing production operations. As a result, a system of work-arounds must be maintained to overcome these deficiencies, forcing both approaches to coexist simultaneously during the transition. Among those we studied, we found this mixed approach never to have been successful. Given the choice, many within the workforce will cling to the old method, never giving the new system a chance to take hold. Thus, despite a strong effort, the system may fail miserably.

A common symptom of these conditions is a widespread disregard for the master production schedule (MPS). In fact, some organizations work around this formal schedule of day-to-day operations so frequently that they have come to see it as a means to satisfy management requirements rather than as a useful tool. Here, it may not even be distributed to production planners; instead, it is set aside to avoid confusing the system of work-arounds that actually drives the factory operations. Since the MPS is a primary output of the production and inventory control system, its ineffectiveness signals the limited degree to which this broader system is actually put to use.

Fortunately, we found that early problems do not always spell disaster for a production and inventory control system. One facility that we visited had initially experienced difficulty, but was able to recover fully through close monitoring and rapid corrective action. In fact, we found key data that was gathered as this process played out to be very useful in its analysis. As illustrated in Figure 5.5, prior to putting in place a new system of control—in this case an MRP II system—the facility's production of successive units tracked along a clearly defined learning curve. At point A, the facility attempted only partially to implement this rigorous production and inventory control system. By hedging its bets, concurrently maintaining its prior system, its operations were met with significant pain, as evidenced by an increase in the time required to manufacture each unit as marked by point B on the chart. After regrouping, a different approach was adopted, one that forced factory-wide discipline in using the new system. It appears that by eliminating the coexistence of two opposing methodologies within the same facility, dramatic reductions in manufacturing cycle times were quickly realized.

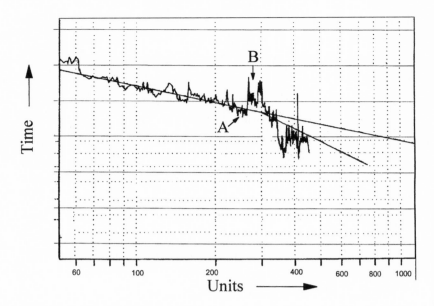

FIGURE 5.5 Learning curve effects.
Source: The Manufacturing Affordability Development Program, Final Report.

The task of implementing a successful production and inventory control system, one with the capability to drive down production variation, should not be trivialized. A specific strategy, a detailed implementation schedule, and a great deal of commitment are all needed. Even with this system in place, rigorous implementation remains essential—without it, all other preparation is for naught. The desire to progress quickly must be restrained, with controls set up ensuring that key enablers such as attaining high inventory accuracies are *always* met prior to proceeding to subsequent implementation phases. With this type of rigor, the chances for success are greatly increased.

When selecting a production and inventory control method, the value of simplicity cannot be overstated. This was clearly demonstrated by those that adopted line-of-sight systems; those that restructured their operations to permit this type of approach reaped great benefits. Yet we also found that even complex operations can benefit from implementing such systems as MRP II in its simplest form. By adopting the hybrid approach discussed earlier in this chapter, fabrication shops can be controlled by line-of-sights approaches, allowing the MRP II system to keep the broader planning activities in lockstep. We found that this approach minimizes the administrative burden of the system while increasing the chance that it will provide real value to the organization.

A facility must also ensure that it has considered its long-term strategy prior to selecting specific tools to accomplish its production and inventory objectives. For instance, we discussed two dramatically different strategies to attaining high inventory accuracies. By selecting the more straightforward solution of implementing strict physical control, a facility can achieve its near-term goals, yet in the process it may be damaging its future ability to continue on this path of improvement. The practice of locking up inventory will do nothing to foster workforce buy-in to the system, a key factor to a number of next-level enablers (to be addressed in subsequent chapters). As will be discussed in Chapter 9, only by working in conjunction with the workforce can workforce discipline truly become instilled. Only with this accomplished will a facility be able to implement the types of tools and practices that can drive the facility to the next level of performance.

CHAPTER 6

The Impact of Production Flow and Quality Enhancements

The application of enhancements to the production line, whether in the form of automated equipment, adjustments to work methods, or increased attention to product quality, has for almost a generation been viewed as the best hope for manufacturing cost reduction within the aircraft industry. These initiatives have been applied toward furthering such goals as shortening manufacturing queue times through the reduction of travel distances and machine setups, optimizing production floor layout, and generally increasing the productivity of the workforce. Together, their desired effect has been to streamline work flow, allowing the overall production process to be performed much more quickly and efficiently.

The direction of this industry in approaching this objective has been highly inconsistent. An early emphasis on quality as the key to controlling production waste gave way to a push for automation as the panacea for all that ails the factory. As the quest for the ultimate automated plant (or "factories of the future") faded in the late 1980s, other approaches began to take hold, each promising the reward of a vast untapped potential for improvement. Many of these latest trends appear to hold genuine opportunity for dramatic improvement; cellular manufacturing, workforce empowerment, SPC, and flow time reduction tools have all been successfully demonstrated elsewhere.[1] Unfortunately, our study showed their results to be mixed. For all of the effort spent pursuing these, very little actual factory-wide savings could be quantified at a majority of the facilities we visited.

While these concepts all appear to be valid, why is it that they did not return the degree of savings that was anticipated? After all, who can doubt that poor quality drives up the cost of manufacturing, causing rework, production delays, and increased inventories? Shouldn't the automation of a process lead to improved control of variation, ultimately reducing manufacturing cycle times, increasing quality, and leading to on-time deliveries? A number of other lean approaches that promise to cut costs through increased factory responsiveness and reduced waste seem to be equally reasonable. Why then have the substantial efforts expended toward these goals not consistently translated into broader savings?

We found that the problem lies not in the merits of each of these tools individually—most are based on sound concepts bolstered with proven examples of success. The problem instead lies in how they are applied. In Chapter 3, we demonstrated that when production flow and quality enhancement tools are implemented discretely, there is no guarantee that any real savings will result. Unless an orderly approach to controlling production and inventory is first adopted, efforts to improve other areas are unlikely to consistently yield factory-wide savings. As with our traffic analogy from Chapter 4, without first ensuring that we have provided an adequate road on which to travel, additional measures to tweak individual sources of variation tend to be ineffective.

Early Attempts to Improve Production Flow

Throughout the 1980s, many sought to develop the means to improve factory productivity by applying various methods designed to increase product quality. The costs associated with quality-related problems were readily apparent: widespread material and labor waste from the associated scrap and rework of completed items, disruption to production flow, and large inventories to cover for low production yields. It was felt that by controlling this widespread waste, unprecedented gains in factory efficiency could be realized. After all, the Japanese automotive industry had demonstrated this to be possible, having yielded dramatic cost gains ostensibly as a result of controlling these problems.

We found that the industry approached this largely through two different means: by striving to enhance the effectiveness of the workforce, and by investing in equipment to automate manufacturing processes. The former was based on the Japanese methods of simplifying and controlling processes through such methods as *kaizen* events and Total Quality Management. The latter was based on an even more direct approach: By employing automation such that manufacturing operations could be repeated with greater precision, less variation would likely be introduced into manufacturing processes. Thus, a major source of day-to-day disruption could be eliminated, leading to streamlined production operations.

Some facilities moved out aggressively to demonstrate these methods. Automated equipment was employed for a range of tasks such as machining complex parts and even the drilling and reaming of close-tolerance fastener holes. Others developed equipment to replace such tedious and error-prone jobs as locating and soldering delicate components to electronics boards.

This permitted a much greater degree of control over these processes, allowing the type of repeatability essential to producing the high degree of quality and reliability demanded of these items. Through the use of this process-targeted automation, human-induced error could be minimized, thus greatly reducing a major factor driving production cost.

At the same time, manufacturing technology was maturing to the point where some believed that this sort of automation could completely take over the factory—thus eliminating human interaction from the manufacturing process altogether. Experiments with this concept ultimately led to the implementation of flexible manufacturing systems (FMSs) (or wholly automated facilities). This signaled to some that the day had come where a "factory of the future" would operate without the problems associated with human intervention. Enabled by various specialized "robots" such as automated storage and retrieval systems (ASRS) for storing and organizing extensive racks of inventory and automated guided vehicles (AGVs) shuttling materials across the factory floor to be automatically loaded into processing machines, these facilities were intended to all but eliminate material shortages and manufacturing errors. A number of companies configured one or more of their shops into these "factories of the future" in the hopes of reaping the huge savings that these systems appeared to promise.

Ironically, very few of these operations still exist today. Late in the 1980s, many began to realize that these plants were actually *not* cost-effective. Several manufacturers recounted to us how this approach had driven them to operate at inefficient rates. The combined effects of the high fixed costs of automation equipment along with the severe lack of factory flexibility that came with using such specialized equipment made it very difficult to consistently optimize production operations. High production rates of similar items were needed to pay for the equipment—a condition that does not often occur in this industry. Consequently, many of these operations have been once again converted, with much of their automated equipment removed or used for other purposes. In some cases, we saw this equipment sitting idle, still incurring costs for periodic maintenance solely for the purpose of preventing this very large investment in facilities and equipment from completely deteriorating.

In retrospect, this move is consistent with the latest direction of the automobile industry. It has been reported that Toyota, one of the most innovative manufacturing companies in the world, is rethinking its use of heavy automation. Toyota was once seen at the leader in high automation with robots performing a wide variety of assembly tasks, but these expensive pieces of machinery are curiously absent from its latest vehicle line (the RAV4). Because of the high maintenance costs and inflexibility associated with highly automated facilities, Toyota no longer sees this as the most cost-effective solution. An approach that more efficiently leverages the strength of its workers appears to be viewed as a more efficient direction.[2]

The Advent of Cellular Manufacturing

As early as the late 1980s, many of these same facilities began to move in a much different direction: They began to implement a productivity improvement technique known as cellular manufacturing. This concept is built on much of what had been learned in the race for factory automation, but with an emphasis on production flexibility. Over recent years it has continued to gain favor, and now appears to be widely viewed as *the* means to provide significant rate and product flexibility at much lower costs than automation.

Cellular manufacturing is defined here as the grouping of people, equipment, and processes for the purpose of performing complete segments of production more efficiently. While traditional manufacturing shops may only perform discrete specialized processes, manufacturing "cells" are intended to perform a complete series of production operations from beginning to end. This can result in the production of an entire component, the manufacture of a finished end product, or the completion of a major severable task within a larger operation.

Each cell consists of a largely self-sufficient collection of the equipment, facilities, and workers needed to support its assigned job. For this reason, they tend to be focused on the production of groups of items with similar processing needs, setup times, fixturing, and inspections. This specialization of the cell's functions permits the minimization of equipment, floor space, and skill mix needed for the workers to perform a complete set of tasks for a large group of items. Thus, a single worker can often perform most or even all of the tasks within the cell. Since this allows workers to be easily shifted from job to job to support varying production demands, the flexibility of manufacturing operations is greatly enhanced. This also simplifies the job of the production planner; production planning need not be as precise since work can be rescheduled with little notice to lend support across a range of factory needs.

With individual workers now responsible for ushering parts through complete phases of production, it is more likely that they will develop a sense of ownership in the work they perform. By eliminating the need for an item to be passed through the hands of multiple operators and material handlers, the root causes of delays and quality discrepancies can be much more readily identified. Instead, this single operator can ensure that his or her part is of the highest quality and is delivered on time, thus preventing many of the problems seen when control over the same part shifts from department to department. This increased visibility into the roadblocks that routinely impede workers' ability to perform this function will place them in a better position to recommend changes to improve the performance of the cells. With some degree of flexibility maintained by allowing these cells to be somewhat reconfigurable, the workers can rapidly influence the improvement process.

In some organizations, these cells are allowed to operate as semiautonomous units. Often with only their input and output controlled by others, they are empowered to varying degrees to make internal adjustments to achieve the expected end result. This opportunity to tweak their operations to suit changing factory conditions has, in some cases, tremendously enhanced their effectiveness in dealing with the production line as a whole. By coordinating more directly with their suppliers and customers, some have developed an increased ability to manage unplanned surges. Their freedom to experiment with different arrangements of manufacturing operations has allowed them to find solutions best suited to their unique environments. This tailored flexibility has allowed some to make deliveries on time consistently despite a number of seemingly disruptive influences driven by factors external to their own operations.

Because of early successes, some have pushed this concept of worker empowerment much further, expecting it to help them attain even greater benefits. In one case, each of the cells in a facility is provided its own financial analyst. As workers develop suggestions for improving product flow, lowering inventory levels, or stocking office supplies and machine tool cutters, they can now develop real-time business cases. If the improvements make sense, management acts quickly and decisively to approve the necessary budget. As a result, these cells have made large changes that would not have been possible without this streamlined source of funding.

Not all examples have been met with success, however. One facility, pleased with its early results, pressed this concept much further by allowing totally self-directed work teams. Here, management oversight had been essentially eliminated, leaving workers to manage all aspects of the cell's operations. At the time of our visit, these cells were experiencing declining productivity. The company was in the midst of investigating the reason for this, but continued to believe that further experimentation with this approach will eventually produce an optimal solution.

Despite some of these problems, it is easy to see why many have turned to cellular manufacturing as a potential answer to the limitations of advanced automation. Since a highly automated factory must be optimized for a set production rate, subsequent fluctuations in actual demand may lead to either suboptimal use of equipment or the buildup of excess inventory—both of which are costly situations. Conversely, a cellular manufacturing approach offers much more flexibility in the use of these machines; its latitude to continuously rearrange work or even reconfigure the shop layout allows factory flow to be optimized based on changes in product mix and production rate.

Much can be learned from the aircraft industry's widespread experimentation with this concept. Lessons demonstrating both success as well as failure across wide-ranging product mixes and complexities lend a unique degree of insight into the value of varying implementation strategies. Armed with this information, we can begin to understand how this approach affects manufacturing cycle time, inventories, variation, and ultimately cost savings.

IMPACT ON IMPROVEMENT METRICS

We found the most consistent contribution of cellular manufacturing to cost savings to be the dramatic reduction in factory dead time, or the time an item spends in shipping, machine queue, storage, or any other activity that adds no real value to the end product. A great deal of this result can be attributed to the process-flow evaluation that is necessary to facilitate a shift to a cellular manufacturing approach. Facilities that evaluate their existing operations to prepare for this shift are often slapped with the realization that sequential operations are significantly dispersed. By merely moving these operations closer together, a dramatic amount of travel time can be eliminated.

This is analogous to a similar type of savings from the practice of process automation. Before automation can be effectively applied, a rigorous analysis of the process, or "pre-automation,"[3] is performed. By shortening the travel distance or number of processes a part has to travel through, or by making the part easy to access or set up for a given operation, a great deal of flow time savings may be possible even before the automation is put in place. Looking at these preparations can reveal that the benefits offered by automation may often really be tied to these preparations rather than the automation itself.

We found a number of cases where this could in fact be shown, that the act of revising the factory layout was what led to the real benefits. Before one plant moved to a cellular plant layout, it found that parts traveled many miles during the course of production. Rearranging equipment, however, reduced travel distances to only a few dozen feet. After the machines were moved closer together, it no longer made sense to shuttle the items they produced back to a central staging area between sequential operations. Consequently, the number of items in circulation around the factory was reduced substantially, thereby lowering overall inventory costs by 45%. Cycle time was also reduced by about the same amount because of this reduced processing time.

Unfortunately, we found that many facilities became sat-isfied all too easily, stopping their pursuit after making only modest gains. One facility was quite proud of a simple factory rearrangement that reduced its product wait time to process-ing time ratio from 40:1 to 9:1. We pointed out that this im-provement still meant that the part spent 90% of its time in the factory waiting to be processed.

Several proceeded much further than these initial steps with the cellular manufacturing concept. By developing quick-change fixtures and simplifying procedures, some facilities were able to reduce the amount of time required to set up a part for processing by as much as 95%—from hours down to minutes. These actions also made these operations much more predictable, reducing cycle time variation dramatically. In some cases, the time—and cost—for setup was sufficiently reduced to make smaller and smaller batch sizes possible. While some sought ultimately to drive this trend to a condi-tion known as "one-piece flow," we only occasionally ob-served examples of this, even in the most advanced of cellular manufacturing applications.

One-piece flow is a technique that strives to limit a fac-tory's operations to producing only what is needed to support current demand levels. The following is the most common in-terpretation of this philosophy that we found used across the aircraft industry: By simplifying setups, the cost of changeovers can be sufficiently controlled to result in an *eco-nomic lot size of one*—where there is no cost impact if produc-tion were shifted each time one ship-set of parts was manufactured (those items needed to support the production needs of the next end item). With this, it is feasible to change over tools or machines to produce different items precisely when they are needed. Under this construct, planning for ma-chine use is greatly simplified; a cell can maintain efficient operations even without the ability to anticipate changes in demand levels. As long as the overall capacity of the cell is not exceeded, operations can be shifted—almost at will—to respond to the specific demands of the day. Many have looked to this approach as a direct means of saving costs: By reducing production lot sizes, inventory levels (especially WIP) are also reduced. This results in the overall cost reduction depicted in Figure 6.1.

Unfortunately, those that have experimented with this concept have shown that there are substantial barriers to implementation. A move to one-piece flow often requires equipment to be selected and the product designed specifically to support this concept. Existing equipment originally intended to support batch operations often cannot be adapted, necessitating costly replacement to support this changed philosophy efficiently. While existing product configurations can sometimes be modified to some small degree, the broader changes needed for dramatic, widespread simplification of setups can

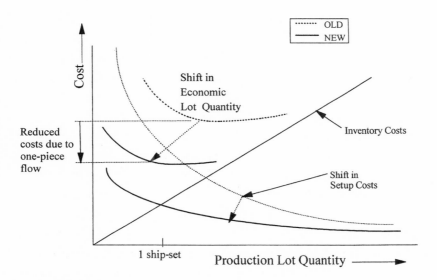

FIGURE 6.1 Impact of reduced setups on manufacturing costs.

be difficult to accommodate (these constraints will be discussed further in Chapter 8). As a result, while many within this industry recognize one-piece flow to be a valid philosophy, they do not see it as practical to set these goals within their operating constraints.

Instead, a number have adopted a less aggressive approach to configuring manufacturing cells for reduced batch size operations. By improving their control over production operations, they can more efficiently plan equipment changeovers, resulting in lower production costs. Often, this lower cost can be directly translated to a higher demand for the cell's products, leading to a better justification for maintaining production in these factories rather than giving it to subcontractors. This consolidation of larger quantities of similar items into these cells results in increased production rates, greatly improving efficiencies. Since higher production demands lead to a more rapid turnover of WIP inventories, even the costs associated with batch-type production can be greatly reduced. This is because the inventory spends less time in the factory accruing inventory costs. Thus, as depicted in Figure 6.2, while an emphasis on reduced setup times may fall short of one-piece flow, we found that it can still allow dramatic gains in cost reduction. Still, it does not offer nearly the flexibility to shift operations to respond to changing demands—a key feature of cellular manufacturing.

We also found that cellular manufacturing has enabled large cost improvements by facilitating the planning and control of factory operations. Breaking a facility into a series of work cells permitted to manage their own inventories and work flows greatly simplifies the function of the production manager. No longer must the factory's production and inventory control system attempt to control work down to the day-to-day level. Instead, its functions can be

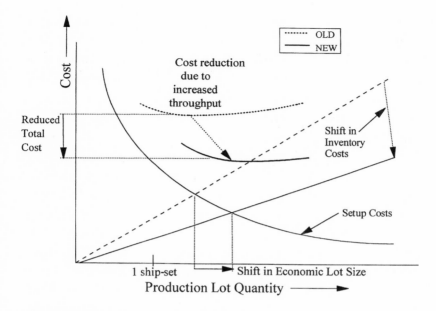

FIGURE 6.2 Impact of increased throughput on manufacturing costs.

concentrated on the much more manageable task of tracking material or work entering or leaving each cell. As a result, data management is greatly simplified, resulting in much greater accuracy. Even more importantly, this simplification greatly increases the system's ease of use, making it much less likely that the workforce will try to work around it. For these reasons, we found that this application of cellular manufacturing had a tremendous impact on the successful implementation of complex production and inventory control systems (as we discussed in Chapter 5 when covering the MRP II and hybrid systems).

While we could not break out the factory-wide cost reduction contributions from improvements in dead time, inventories, WIP, or the improved production flow control offered by cellular manufacturing, we did find indications that their effects are large. Some shops were able to reduce floor space by 40%, scrap rates by 90%, and inventory costs by up to 80%. In the most effective applications, we also saw a dramatic reduction in both cycle time and cycle time variation, as well as a drop in processing costs of up to 50%. A number of intangible improvements were clearly visible, such as gains in customer satisfaction and product quality.

While these examples allow us to begin to bound the savings potential offered by cellular manufacturing, it is important to note that they do not represent the norm within the aircraft industry. Although the facilities we visited were often able to demonstrate sometimes impressive gains, these were typically very localized in nature, rarely leading to broad-based savings. As was briefly discussed in Chapter 3, only a few facilities had achieved significant factory-wide improvement. We found this to occur primarily because important precursors had not first been implemented (i.e., control of production and inventories). Yet, as we investigated further, we found that a more specific pattern emerged. We were able to identify key aspects to implementation that repeatedly led to success:

■ The establishment of meaningful, facility-based goals,
■ A rigorous maintenance program, and
■ The existence of a clearly identified focus.

Because of their significance, we will discuss each of these in turn.

THE ESTABLISHMENT OF MEANINGFUL GOALS

It was not uncommon to see manufacturing cells that had been truly successful in accomplishing their original goals only to realize little or no impact on the end cost of their products. We found other shops that had been touted as "cellular success stories," but actually did not hold to this concept at all. Closer scrutiny revealed them to be clever adaptations of process automation that simply enhanced old-style process-based manufacturing concepts. In one such case, claims of large increases in productivity did not hold up when viewed from a broader perspective. Instead, it turned out that the entire cost savings attributed to this "improvement" was likely offset by the cost of excess inventory produced to maintain so-called "cost-effective rates."

We include this observation to underscore an important lesson: Successful implementation of cellular manufacturing requires a solid plan that *supports the overall goals of the organization*. Without a detailed understanding of this concept and what it is intended to do, it is very difficult to develop meaningful, achievable goals. If it can't be seen how these goals will have a broad-based effect on the organization once they are achieved, it is unlikely that the effort had been fully thought out.

While applying this type of logic when embarking on any improvement strategy may seem intuitive, it is surprisingly often neglected. An investment in automated equipment to speed up production at a given workstation makes little sense, for example, when the process in question is not pacing the line. The application of improvements to a process that doesn't present a cycle time, inventory, or variation problem to a facility may be very costly. In order to make these investments pay for themselves, machine utilization must be kept high; without sufficient demand for this processing, the cell may be led to produce piles of unneeded inventory. Facilities that have learned painful lessons refer to this counterproductive flow of work as either "feeding the monument" or "feeding the elephant."

Only those facilities that first identified their real needs have been able to leverage production flow improvements into dramatic, facility-wide savings. A solid application of the foundation level enablers of Chapter 3 is critical; only when these first-order sources of variation are successfully managed can a plant gain a sufficient degree of insight into its shortfalls. As shown in the highway analogy of Chapter 4, only once a sufficient road is provided can further efforts to smooth traffic flow be fruitful. With a firm degree of production and inventory control established, it now becomes possible to identify the true bottleneck operations, allowing genuine improvement opportunities to be targeted. Automation can be applied as appropriate where its ability to improve repeatability adds real value, such as addressing such limiting factors as poor quality or low process yields. Similarly, cellular manufacturing and its enabling tools and practices can be targeted to result in a facility-wide impact. When implemented without these broader considerations, facilities instead tend to create "islands of excellence," with results more useful as marketing anecdotes rather than as actual cost savings.

———

While the concept of cellular manufacturing seems to signal a distinct move away from the use of automation as a production flow improvement tool, automation actually plays an integral role in some of the more successful examples of cellular manufacturing that we have seen. By carefully selecting the equipment used to facilitate the operations of a cell, the workforce and machinery can complement each other to result in a very powerful system. The selection of flexible equipment can allow a worker to perform multiple tasks without repeated setups. Individuals who were once focused on highly specialized, singular tasks can now be trained to perform several functions within the cell. Often they can accomplish this with less training than was needed to perform each individually. This ability to perform a range of tasks can provide this same workforce the flexibility to process a somewhat diverse group of items, increasing throughput by optimizing the use of people and equipment.

Those facilities that have successfully melded automation into their manufacturing cells have done so by recognizing not only the strengths, but also the limitations of this approach. Too often, automation has created more problems than it has resolved. In order to prevent this type of investment from introducing new variation into manufacturing operations, a strong focus in preventive maintenance is needed.

THE IMPORTANCE OF PREVENTIVE MAINTENANCE

In order to prevent unscheduled delays to production operations, the dependability of a cell's equipment is critical. With the roles of both the cell's equipment and its personnel integral to its performance, a misstep by either can cause severe disruption to operations. As a result, it becomes important to closely manage both. Procedures must be developed to minimize human error, and supporting equipment must be maintained such that its failure is rare.

We observed two techniques that appeared to be particularly instrumental in achieving these goals. A method similar to one known as *poka-yoke* or "foolproof fixturing" has been used to help prevent errors in machine setup, ensuring that parts cannot be loaded in an incorrect manner.[4] Where this has been applied, operator-induced processing problems have been greatly reduced. Another technique, known as Total Preventive Maintenance (TPM), is intended to minimize unscheduled downtime of production machinery. This concept was created by General Electric in the 1950s and further proven out in Japan as an integral part of the emerging Just-in-Time approach.[5] When successfully implemented, it has been highly effective in minimizing unanticipated equipment failures, thus increasing the overall dependability of the cell.

The TPM concept embraces not only the concept of routine maintenance, but also that of *predictive* maintenance. While routine maintenance must be performed to fulfill the basic servicing needs of the cell's equipment (such as checking and filling lubrication oil receptacles and identifying warning signs of impending machine trouble), predictive maintenance is intended to anticipate and actively manage the time that this equipment would otherwise be down for unscheduled, corrective maintenance. Feedback from routine maintenance as well as SPC data can reveal minute changes in equipment performance that can be used to predict the need for maintenance actions well before processing capabilities are degraded to unacceptable levels. With this information, requirements for special maintenance actions can be forecasted and scheduled with sufficient lead time such that work-arounds can be planned. Often, equipment is scheduled to build temporary buffer stocks of inventory to support production operations during the period in which this equipment maintenance is performed.

We found this maintenance approach to hold great promise in managing this source of variation without introducing excessive inventory costs. When rigorously implemented, this has led to the dramatic reduction of equipment downtime, while trimming the requirement for offsetting inventories to bare minimum levels. In one case, confidence in this prediction of maintenance needs was so high that inventories were built up only to the extent that they could be depleted within two hours of the completion of machine maintenance. Using this approach, equipment can be tightly scheduled without the need to maintain long-term buffer inventories.

THE IMPORTANCE OF A CLEARLY DEFINED FOCUS

Regardless of the contributions made by other enabling factors, the selection of the specific work to be performed by individual manufacturing cells is often key to permitting their success. It can be reasoned that the more a cell's mission can be narrowed, the smaller the range in capabilities from equipment and worker skills that is needed. Cells assigned to manufacturing items with very similar physical and manufacturing characteristics can rely on fewer machines, skills, and work space. These similarities can result in more common operations, reducing the complexity of machine setups when changing between the production of dissimilar items. Greater commonality in the materials needed to support production operations can streamline material management and supplier coordination functions. In all, a narrow, clearly defined focus can greatly simplify the operations of a cell, leaving less room for variation to creep into the processes.

In order to enable this type of alignment, a structured approach is critical to efficiently and effectively organize the vast number of items and functions found within a complex production operation. As was discussed earlier, a strong degree of production and inventory control forms an essential part of this approach. Once this is achieved, however, a means to further understand and categorize production operations is needed. We found one system in particular that represents an effective approach to driving this change: the use of a concept known as group technologies.

Group Technologies as an Improvement Enabler

Group technologies is a concept designed to take advantage of the similarities shared by individual items. Using a rigorous methodology, those items that share design or processing characteristics are segregated into part families. They may or may not look similar to one another; what is important is that they share "common materials, tooling, setup procedures, labor skills, cycle time and, especially, work flow or [process] routings."[6] The development of these families can provide substantial value, allowing commonality between items to be leveraged to simplify design and manufacturing operations.

The concept of group technologies has been applied to such tasks as simplying the job of engineering new products.[7] With access to the list of all available items sharing common materials, tolerances, and key design details, a design engineer has the visibility necessary to utilize or adapt existing items instead of developing them from scratch. This innovation has been responsible for saving countless design hours, as well as reducing the quantity of different parts demanded from the manufacturing organization.

This same concept has been used to facilitate the simplification of production operations within complex facilities. By grouping together items that share common manufacturing traits, routings can be greatly simplified, reducing the complexity of factory management. The insight gained from this approach is especially powerful as a precursor to cellular manufacturing, since groups of items with common routings form natural targets for manufacturing cells. By first gaining this type of insight for all items produced in a facility, manufacturing cells can be set up with the greatest possible throughput, greatly enhancing efficiencies.

There are indications that the aerospace industry began considering vaguely similar concepts even before World War II.[8] However, for reasons unknown, efforts to broadly capitalize on processing commonalities never fully materialized. For decades very little was written on this subject in the English-speaking world, with British references surfacing only decades later.[9] Late in the 1970s, the Japanese revived this concept, adapting it for use in their electronics factories.[10] Only much more recently have U.S. companies begun to grasp the significance of this approach.

In the 1980s, group technologies became a hot new approach, adopted by many as a primary means to better organize production operations. At this time, the prevailing implementation strategy was to leverage the capabilities of computers to sort items into product families. Unfortunately, this approach was not as straightforward as was originally envisioned; the effort to code these items to permit the sorting process could take years to complete.[11] Our own review uncovered one facility still actively pursuing this approach, attempting to encode hundreds of items using dozens of characters to represent key processing requirements. Like many before them, they eventually abandoned the idea in favor of other approaches with the potential for more immediate payoff. In retrospect, an approach that requires the use of such a complex coding system may have undermined the underlying purpose of the concept.

One Japanese company, determined to break through the sheer complexity of the parts lists for their various products, developed an innovative approach to streamline the process. By subdividing the bills of materials into small, manageable groups based on where the items are used in the assembly process, similar product-grouped bills of materials could be compared side by side. Common parts—or those that could be made to be common—could then be much more easily identified. Eventually, the factory was realigned

to take advantage of these common processings and routings, with great success.[12]

Our study revealed similar methodologies used to identify
process-based groups of items within the aircraft industry. Organized examinations of route sheets, bills of materials, and
demand histories were performed, and items found to follow
similar processing flows became a "family." Teams of workers
and support engineers could tailor these reviews to different
degrees of fidelity based on the needs of the organization. By
using this information to rearrange the factory layout, these
families of items could be aligned into common manufacturing cells. What is especially noteworthy is that this approach
could be completed within a period of months (rather than
years), and, where rigorously implemented, it acted to enable
the most successful examples of cellular manufacturing that
we observed.

This type of detailed assessment of factory capabilities and
processing requirements can lead to new ways of viewing factory layout and production flow. Armed with a better understanding of the specific capabilities of the factory to produce
well-defined families of parts, an organization will be much
better equipped to make changes allowing overall factory flow
to be vastly improved.

The Role of Quality Improvement

Many across this industry have long looked to initiatives designed to control product quality as a precursor to reduced
production costs. This premise can be logically argued: Reducing the number of product defects should bring about
concomitant reductions in rework, scrap, work-arounds, and
factory disruption. As a result, cycle times and buffer stocks
held in inventory can be lowered, as well as overall factory
variation related to late deliveries. Our observations re-

vealed, however, that this is not always the result—an effective, formal quality control system has not necessarily led directly to success.

Poor quality, much like weaknesses in production and inventory control, can lead to variations in production, potentially impairing the ability of an organization to effect meaningful improvement. Yet, as an almost ubiquitous source of irritation, it can easily be interpreted as *the* cause of disruption, often associated with such broadly based evils as high buffer inventories, late parts deliveries, and out-of-station work. We found that this assumption can not possibly be true, since efforts to improve quality in isolation from other basic enablers do not necessarily lead to a dramatic improvement against these cost drivers. Instead, as we found with the application of production flow improvements, only when efforts address genuine bottlenecks—or those constraints that pace factory flow—do strong quality systems result in dramatic cost improvements.[13]

As with the other areas we have discussed, the need to first achieve a strong degree of production and inventory control is paramount to realizing a substantial return on quality control initiatives. Only when these more basic factors are first addressed (as outlined earlier in this book) can a focus be developed that will allow a consistent contribution to reducing overall production variation. Without first gaining control of these first-order enablers, the true degree of disruption caused by poor quality cannot be readily seen. An unwarranted degree of emphasis in this area can prevent a facility from targeting the true underlying problems.

Besides an inconsistent emphasis on these underlying enablers, we also found this limited impact from quality initiatives to be tied to an ineffectiveness of the techniques implemented. Systems such as "cost of quality," intended to provide information on the effects of quality on the facility, were rarely found to focus on the most significant drivers. They instead attempt to quantify the much smaller direct costs associated with rework or scrap of defective items, while ignoring the much larger effects of late deliveries on the production line. Furthermore, widespread flaws in data entry typically render any credible attempts at analysis almost impossible. At one facility, the same defect is recorded differently each time it is found, often at different levels of the bill of materials—sometimes at the component level, other times at the system level. As a result, the system can not accurately highlight the existence of a recurring defect, nor can it effectively identify true bottlenecks. It is hard to see how such a flawed system could be leveraged for any meaningful benefit.

While, in most cases, attempts to implement SPC were equally disappointing, we observed two facilities that had been successful in their implementation. By studying the impact of quality problems on the entire flow of their production lines, they were able to prioritize efforts on those processes with the potential to impact operating costs significantly. As a result, their programs had substantial, rapid impact, allowing them to gain momentum for broader implementation.

In this industry, poor quality is unacceptable. It can lead to serious flight safety implications, and therefore requires rigorous methods to be implemented for assurance. As a result, the products delivered to a customer are consistently of a very high quality—but, *the cost of achieving this can be tremendous.* It appeared to us that approaches to reduce this cost are only consistently effective when performed in conjunction within an overarching structured approach to improvement.

The Promise Held by Contemporary Improvement Practices

It now appears that much of the aircraft industry is settling on a consistent method for improving production flow. A focus on cellular manufacturing as a core element allows workforce flexibility and factory automation to work together not only to provide greater control over production flow, but also to do so at an increased rate and with greater productivity. Previously attempted methods such as continuous improvement can now take hold; after providing the means for workers to identify logical improvements to production flow, this approach can finally be used in the manner in which it was intended.

It must be reemphasized that in order to make this methodology work, a logical structured approach must be used. As soon as the core enablers are controlled, the stage is set to allow bottlenecks to become visible. Once this is done, methods such as group technologies can provide a systematic, logical methodology to view a system that is otherwise too complex for the detailed degree of analysis needed.

Much as we cautioned in the previous chapter, it is important to note that a facility's long-term strategy must be considered when embarking on approaches such as cellular manufacturing. Earlier we noted that progress can be made even where barriers to one-piece flow exist (such as the cost of upgrading production equipment and product designs intended to support batch operations). While we argued that short-term savings are possible even despite these contraints, we do not feel that this should be viewed as the endpoint objective. Without a concurrent emphasis on removing these constraints, it will likely be difficult to achieve further gains, or even to remain efficient if the demand for individual items shifts significantly. If, however, a facility takes on the mindset that new product designs and factory upgrades will be aimed at facilitating reduced setup times, and current equipment will be continually reassessed with this same emphasis, the stage will be set for further increases in savings and enhanced factory flexibility.

While many facilities have adopted the tools and practices

discussed within this chapter, few have succeeded. We have visited a number that moved out aggressively, only to stumble and ultimately fail in their efforts. It is easy to fall prey to, as one manager put it, "the need to err on the side of action." Without an organized methodology, "the need to err" will become precisely this; the organization will have to fail to realize the flaws in its approach.

Those that have succeeded have demonstrated the great possibilities of these tools and practices. Not only can they make dramatic savings possible in a relatively short time frame, but they can transform the way a factory does business, allowing it to achieve a level of productivity and smooth flow never before imagined possible. When implemented within the context of variation management, they can open the doors for higher levels of savings.

CHAPTER 7

Improved Supplier Responsiveness

As was briefly discussed in Chapter 4, a key benefit of variation management is its tremendous potential to improve the performance of a plant's suppliers. This is of critical importance, especially within the aerospace industry where suppliers contribute as much as 70% of the value to the end product.[1] Because of this broad exposure, inefficiencies impacting the cost of delivered goods can have a large effect on the cost of the end product. Furthermore, late deliveries and poor quality can substantially disrupt factory operations, thus driving up such key factors as production cycle times, inventories, and schedule variation. In fact, we found that it is often because of the latter effect—the highly visible impact to their own plants' operations—that many have begun to focus attention on supplier base improvement initiatives.

Despite this attention, we found supplier problems to be the most frequently cited concern across the aircraft industry. Late matériel deliveries routinely disrupt factory operations, leading to stock-outs that can impact activities from component fabrication to final assembly. When this prevents items from being manufactured on time to support subsequent production operations, minor delays can snowball into widespread disruptions and work-around activity. Even delays on individual items have the potential to cause a domino effect across the factory.

We found that this industry tends to deal with this external source of variation in much the same manner it deals with variation that originates internally: by taking measures aimed at mitigating its effects, usually through the use of work-arounds. For example, we found that a number of manufacturers maintain in-house inventories of supplied goods, clearly in an attempt to insulate themselves from the effects of a less-than-dependable supply chain. Others have opted against this practice; because of their broad exposure, the level of inventory necessary to gain a sufficiently improved degree of confidence is often cost-prohibitive. Instead, they appear to have adjusted production schedules to account for delivery uncertainties.

In an attempt to attack the problem at its source, much of the industry has begun to take a hard line against deficient deliverers by holding suppliers solely responsible for any delays or deficiencies, and either penalizing or dropping those that do not consistently live up to expectations. Preferred supplier programs are widely based on this philosophy, frequently applying contractual stipulations that require the application of specific improvement practices, such as Statistical Process Control (ironically, many of these are commonly not even in place within the customer's own facilities). Suppliers run the risk of losing this business unless costs are controlled, quality is high, deliveries are on time, *and* these practices are strongly implemented.

Still others have come to believe that a technology-based solution offers the best course to improvement. By developing the means to make real-time factory planning information available to suppliers through a web of electronic links, they hope to greatly diminish forecasting problems. After all, if large gains have been possible by providing this type of information within a factory (as shown in the previous chapter), why not extend this same type of tool to the supply chain? If this real-time information were made available to their suppliers, they could anticipate a factory's changing needs, enabling an efficient response.

Unfortunately, the solution isn't quite this straightforward. Many barriers prevent the widespread use of electronic data interchange (or EDI) that would serve as an enabler for this approach. The equipment and software for such systems are far from standardized, making the setup of such a system far from trivial. Many suppliers simply do not have the resources for such technology. For third- or fourth-tier suppliers, the purchase of the enabling computer equipment and the necessary support personnel may be beyond their means. Even Wal-Mart, a leader in EDI commerce, has found this to be the case; of its approximately 50,000 suppliers, only around 5,000 are electronically hooked up.[2] And finally, many manufacturers are reluctant (and often unable) to give outsiders access to the types of sensitive information concerning production rates, inventory levels, and delivery dates that would be needed.

Because of the lack of success to date with many of these approaches, some have begun to question whether any near-term solution is practical. Recent studies have found this problem to be too complex, too difficult to solve with a single, direct method. Instead, some have called for an emphasis on developing more producible products and on driving an overall change to the business environment, solutions that defer improvement until many years into the future—if ever. In this book, we take a far different viewpoint: that by leveraging the principle of variation management, rapid, major improvement *is* possible—even for existing product lines.

Variation in Supplier Forecasts

We found that with many suppliers, the root of the problem lies in their inability to forecast the constant changes in delivery requirements for their products. These facilities must optimize their own manufacturing operations in order to satisfy the highly variable needs of their customers. Despite this lack of visibility, they must still operate efficiently, maintaining solid performance against two seemingly contradictory demands: low-cost production, and the ability to adjust operations readily to respond to their customers' ever-changing needs. Their dilemma is analogous to the highway situation identified earlier: Unable to anticipate changes in the road ahead, they are forced to react to disruptive influences only as they present themselves.

During our review of a number of these customer/supplier relationships, we found that this indeed is a major cause for both delivery and quality problems. Unable to level-load their facilities because of this demand uncertainty, and unable to stockpile finished goods because of severe cost implications, suppliers are left with little latitude for maneuvering. They may be forced to operate in inefficient modes, continually readjusting production rates, building and depleting inventories. Because of their inability to control costs in this environment, profit margins are likely squeezed in order to respond to increasing pricing pressures.[3] It is not hard to understand why so many suppliers have opted to exit the aerospace industry altogether in favor of more predictable markets.[4]

To make matters worse, these same variations in demand ripple through the supply chain, ultimately impacting the purchase of the raw materials needed to produce these products. Because of the competition for limited supplies of such items as castings, forgings, and extrusions, manufacturers must often place orders many months prior to the time they are actually needed. Because of the wide swings in actual demand for products, smaller suppliers that depend on these costly materials may be forced to order large quantities just in case, and may not be able to turn them into finished goods rapidly. The costs of holding these inventories must clearly be passed on to their customers, ultimately to be included in the price of the final product.

Figure 7.1 depicts the variation in monthly demand that might be seen from a supplier's customer base for an individual item. As shown here, demand levels can vary significantly from month to month, appearing largely random in nature. Commonly used forecasting methods are ineffective in this situation. As can be seen in the figure, a prediction based on average trends may not even remotely approximate the actual peaks and valleys in demand that must be satisfied. Thus, it is almost impossible to forecast what must be produced during a given time frame, and the supplier is left with no means to accurately schedule the loading of its factory.

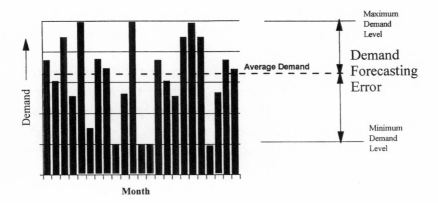

FIGURE 7.1   Notional variation in demand for an individual item.

Traditional Responses to Demand Variation

Despite the fact that there exists no precise means of forecasting these spikes in demand, the supplier must develop a means to respond to them. As was noted earlier, this is often through one of two means: either through the use of strategic inventory stores, or by maintaining some increment of excess factory capacity so that parts can be produced when needed. Each approach is based on a common theme: The capability to respond is provided by holding some resource, either capacity or matériel, in reserve. Without this, the supplier would not be in a position to respond to the customer on time. This could jeopardize its relative standing as a preferred supplier—a factor that is especially critical during this time of downsizing.

Both of these approaches are shown in Figure 7.2. The dotted line crossing these graphs is intended to depict the level of capability that must be maintained to meet this variation in demand. As discussed in Chapter 4, it is very costly to maintain unutilized capacity or inventories to protect against these demand spikes. The costs of holding materials, tooling, facilities, equipment, and personnel to maintain a capacity reserve can be very large. Yet, the implications of not providing items to the customer on time are usually sufficient to warrant these measures.

Despite this, many still end up missing the mark, finding it difficult to fill orders on time. Without the ability to forecast future demand spikes, they can only roughly estimate necessary reserve capabilities based on past experience. Without a more precise forecasting method, it is often found that the risk that orders will not be filled on time cannot be completely mitigated. What is clearly needed is either a means to forecast these individual spikes in demand or the means to eliminate them altogether.

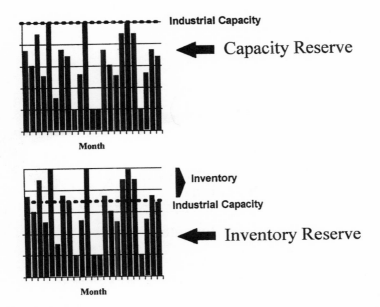

FIGURE 7.2 Manufacturing capacity versus inventory alternatives to supplier responsiveness.

Source: Based on a similar figure posted on the web site mantech.iitri.org/PUBS/ DMC98Pro/Index.shtml as of December 21, 1999.

Customer-Led Improvement

From our discussions to this point, we have seen that the internal effort of a facility to improve its own operating efficiency can in itself make a dramatic impact on its suppliers' support. By first leveraging the foundation building blocks of variation management (depicted in Figure 7.3), a customer can make internal adjustments that can serve to drive its suppliers' ability to improve. As discussed in Chapter 5, by better managing inventory and manufacturing operations, factory variation can be dramatically reduced. Thus, the ability to forecast material needs is greatly enhanced; planners can now minimize the type of expediting operations that have traditionally placed unplanned demand spikes on their suppliers.

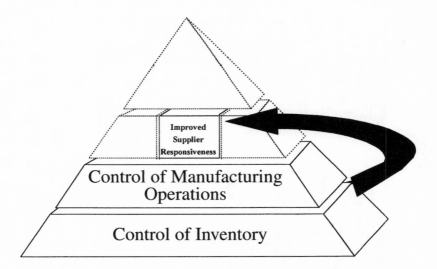

FIGURE 7.3 Foundation building blocks of variation management.
Source: Based on a similar figure from *The Manufacturing Affordability Development Program*, Final Report.

Highly credible delivery schedules can now be developed much further in advance, allowing the supplier base to utilize their resources much more efficiently.

Without this as an enabler, customer-driven supplier improvement has often not been met with much in the way of tangible results. For instance, while many facilities have attempted to reduce their broad supplier bases in the hopes of developing a stronger relationship with those that remain (as seen in the comparability matrix of Figure 3.3), this action by itself had little or no correlation with improvement against our three metrics. However, when coupled with improvements aimed at reducing internal factory variation, these efforts consistently led to dramatic results. Ultimately, by becoming better customers, these factories set the stage for streamlined operations within their suppliers' facilities. This underlying supplier-customer relationship appears to be an iterative process, as depicted by Figure 7.4.

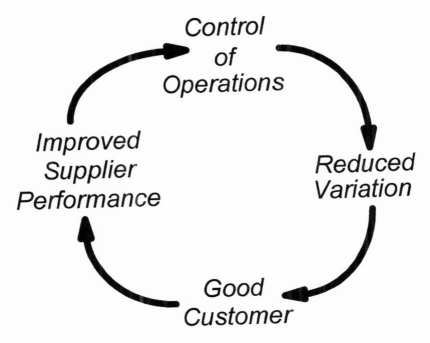

FIGURE 7.4 Supplier-customer loop.
Source: The Manufacturing Affordability Development Program, Final Report.

Unfortunately, we observed few examples where facilities were successful in reducing their internal variation to a sufficient degree to enable this to occur. As was discussed in Chapter 3, we found only a handful of facilities that have instituted the necessary measures. Consequently, most are not yet in a position to facilitate this type of solution.

We must acknowledge that the process of eliminating supplier variation can be viewed differently depending on one's perspective. Figure 7.4 can be used to illustrate this point: Since there is no distinct starting point, the best place to begin this ongoing process can be argued forever—much like the age-old chicken-or-the-egg scenario. While our discussions to this point have shown that it is possible for the customer to take the lead, many of these customers remain convinced that the supplier should also take action. Yet, without controlling the root cause for the supplier disruption—demand variation—how can these suppliers achieve the degree of change that can make a difference?

Managing Variation through Group Technologies

We found that to resolve this dilemma the issue must be addressed from a different angle altogether: Rather than developing the best means to react to a customer's demand spikes, what if a supplier could forge a business relationship that would serve to level out their effects? We found that this is, in fact, possible if each could simply adjust the way that it manages the manufacturing of individual items. In Figure 7.5 we see that, when viewed as separate entities, the demand for individual items may be sporadic and highly variable. As we have already discussed, this can be difficult to deal with. Now, suppose that these individual demands can be merged into a single group, as depicted in the figure. As shown, this will lead to a much smoother distribution—much less severe peaks and valleys. As more and more elements are added, the distribution for the combined group becomes smoother and smoother, ultimately leading the spikes to wash out altogether.

As you can see, while the distribution flattens out considerably, an even more important effect occurs. As more and more items are added to this group, the outlying points of the distribution move closer to their average. As a result, while it remains impossible to predict reliably the variation in demand for a single item, we can now closely approximate the demand for the group as a whole.

While this makes for an interesting theoretical discussion,

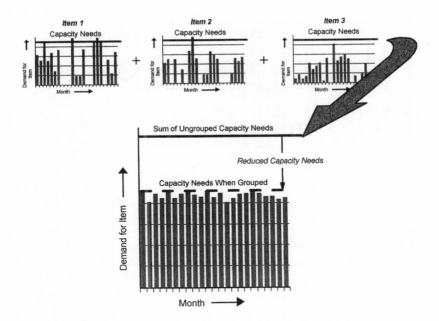

FIGURE 7.5 Flattening of demand through grouped purchasing.
Source: A similar figure is published on the web site www.dscr.dla.mil as of May 23, 1998.

what does it mean to us in real life? To answer this, we must go back to our discussion of the concept of group technologies from Chapter 6. By determining which items share common manufacturing steps, a facility can identify those that could be processed together as larger groups and then plan the purchasing of these items in such a manner as to wash out much of the adverse effect to its suppliers' production operations. Their production capabilities can be utilized more efficiently, allowing products to be delivered on time without requiring costly increases in either plant capacity or inventory levels. Thus, dramatic improvement *is* possible despite the fact that the facility may have done nothing to control its demand variations for individual items.

This effect can be proven statistically. In Figure 7.6, we see how a demand distribution may be seen for a single item.

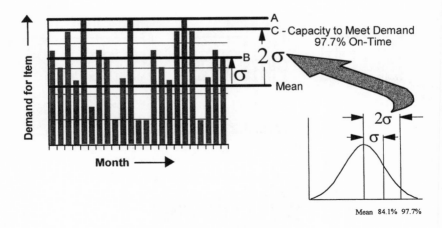

FIGURE 7.6 Levels of responsiveness for supplied parts.
Source: A similar figure is published on the web site www.dscr.dla.mil as of May 23, 1998.

We can determine the mean as well as the standard deviation (σ) of this distribution. Suppose we draw a line across the distribution that touches the top of the highest spike in demand (line A). This line also corresponds to the level of capability that must be maintained at all times (in either excess capacity or inventories) if a facility were always to provide these items on time to their customer. Suppose now that the same facility is only really meeting this demand 85% of the time. This can be approximated statistically by adding one standard deviation to the mean (actually corresponding to an 84.1% probability). This is illustrated by line B. Finally, suppose the facility would like to improve their ability to meet demand to 98% of the time. Line C pictorially illustrates the level that would have to be achieved, corresponding to the value represented by the mean plus two standard deviations.

We found that a series of equations can describe the production capabilities needed for these different levels of responsiveness. Suppose that we combine a number of demand distributions as described earlier. How do we measure the overall effect? The standard deviations for each of these independent, normal distributions can be added using the following relationship:[5]

$$\sigma^2 = \sigma_1^2 + \sigma_2^2 + \sigma_3^2 + \ldots \sigma_n^2$$

For simplicity, the average standard deviation of this distribution can be boiled down to the following (assuming equal standard deviations and means):

$$\sigma = \frac{\sigma}{\sqrt{n}}$$

With these equations, we have the tools to analytically predict the level of improvement offered by utilizing the grouping method described earlier.

In order to illustrate their use, let us suppose that a facility would like to improve its on-time deliveries from 85% to 98% by using this approach. What savings can be expected? The straight horizontal line in Figure 7.7 represents the production capacity for a hypothetical item where the current responsiveness is approximately 85%. This line remains constant as we move to the right, since it depicts a baseline situation where suppliers are not leveraged to apply this grouping technique. The curve on this chart represents the capability needed to deliver at a 98% responsiveness rate. We can see that when our grouping concept is not used (depicted on the left where the horizontal scale corresponds to a group size of 1), the production capability necessary to achieve this increased responsiveness is much higher—as should be anticipated. As we move to the right (corresponding to an increased application of our grouping concept), the capability required to achieve the desired degree of improvement in on-time

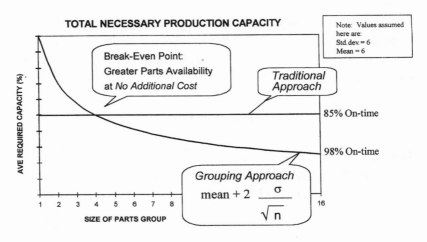

FIGURE 7.7 Analytical payoff of demand grouping.
Source: As published on the web site www.dscr.dla.mil as of May 23, 1998.

customer delivery drops dramatically. From this, we see that it is possible very quickly to reach the point where a higher responsiveness can be achieved with less capability on hand—*and ultimately at a lower cost!*

What does this mean? Production management can be greatly simplified. Managers can now much more accurately predict actual demands, not for individual items, but more importantly *the demands affecting groups of items on common production lines.* This truly provides an agile manufacturing base, one that can respond well to smooth customer demand as well as to production surges.

APPLICATION OF GROUP TECHNOLOGIES METHODS

In implementing this approach, suppose that a customer conducts an evaluation of the overall workload that it will turn over to its supplier base. The customer can break the items that it will ask them to produce into subgroups based on common manufacturing characteristics. Perhaps the company can reach favorable purchasing agreements with these suppliers in exchange for committing to them its demand for these items. If more and more of the purchasing is aligned to allow like components to be produced together, costs should drop while responsiveness skyrockets.

Since some of the effects of this approach can be predicted, both the customer and the supplier can have specific insight into the gains that can be expected even before committing to any changes. The number of items in a group can be tailored based on the level of capacity that a vendor has available, and the magnitude of savings can be anticipated. This should reduce the perceived level of risk, permitting all parties to negotiate agreements based on predictable outcomes. Clearly, this helps to overcome an important barrier to initiating these changes, and provides the baseline for developing a detailed yardstick against which success can be measured.

By incorporating this as part of the make-buy decision process, decisions on when and how to outsource work can now aim at enhancing the flexibility of the entire manufacturing base. By leveraging this grouping approach to optimize both the performance of its own shops as well as that of the suppliers on which they must depend, a facility can improve the efficiency of its entire manufacturing system.

This same approach can be a powerful means of improving the availability of long-lead materials. This is often a critical area; in order to prevent shortfalls, some have resigned themselves to prepositioning stockpiles of materials to overcome their long lead times. Instead, by aligning the purchase of items by groups based on the common use of such materials as castings and forgings, an ongoing stream of materials can be made available without the use of extraordinary measures. Materials can then be ordered based on the level production needs of the group as a whole, with the decision on how to specifically allocate them deferred until actual orders arrive.

⎯⎯▲⎯⎯

Now, suppose that you are a supplier responding to a customer that has taken none of these measures. By adopting certain aspects of this approach, supplier organizations can still achieve a great deal of improvement. While a supplier has little control over the operations of its customer, it does have some degree of control over the type of work it seeks to perform. By pursuing the type of work that readily fits into manufacturing groups, a vendor can structure manufacturing lines to effectively flatten the demand spikes. Long-term planning can focus on capacity and materials usage rather than on the production of specific orders from customers; only short-term planning would be focused on the trade-off of specific customer orders to fill the supplier's capacity. In principle, this could lead to enhanced customer responsiveness and reduced costs, which should allow the vendor to become sufficiently attractive to customers to permit it to pick and choose the type of items it will produce, thus further enabling this type of factory alignment.

In essence, a supplier that does business in this manner is leveraging the concept of group technologies much as may be done by its customers. It really is just the next logical step beyond such approaches as cellular manufacturing that we discussed in Chapter 6—applications that we have seen successfully accomplished within this industry. To illustrate this point, we turn to the machine shop run by the teammate we identified in Chapter 3.

After our study, Don Garrity acted on his new appreciation for the importance of group technologies as he enhanced the efficiency of his facility's operations. He did this by carefully targeting his make-buy decisions and his pursuit of new business to take best advantage of the skills and equipment within his shop.

Much of the efficiency of Don's operations is derived from their flexibility. His machinists are multitasked; they can thus shift from machining parts on a lathe to loading parts on multiaxis machines, again returning to the lathe while waiting for these automated operations to be completed. By seeking to produce items that can take advantage of this flexibility, he enabled his team to readily adjust to wide swings in demand for individual items, maintaining an overall level loading of his shop's activities.

This approach also served to protect his operations from the boom-or-bust cycles seen by the industries to which he provides products. By pursuing business with diversification in mind—producing not only for automotive and aerospace applications, but also items for material handling equipment and faucet manufacturers—he has managed to better insulate himself from these external sources of disruption With his current flexible approach, he can readily adjust his shop to account for surges in demand for each of these individual items, virtually eliminating expediting and its attendant quality and delivery problems.

The results of this approach speak for themselves: For the past three years his shop has been 100% defect-free and 100% on-time for aerospace production parts. This led one of his customers—a major aerospace company—to rank his operation as its number-one supplier, nominating him for a prestigious business award. Even more noteworthy was the subsequent reaction of one of his competitors for this award; after this major company reviewed his facility he was quickly chosen as one of the company's suppliers.

———

The methods that we have discussed within this chapter illustrate the versatility of the principle of variation management. We have seen that this same principle applies to all types of facilities, each responding to different sources and types of variation. Regardless of whether this approach is implemented by a supplier or by its customer, this principle is a key to dramatic cost reduction. Contrary to what may be intuitively believed, we have seen that a customer has a great deal of control over its suppliers' performance. We have also seen that a supplier has a great deal of control as well; while it does not have the power to improve the causes of variation within a customer's plant, it is not powerless to address the effects. By leveraging an understanding of this principle, a vendor can adapt its facility to better respond to the stresses of the external environment. By doing this, it can continue to control its own destiny, even despite seemingly unreasonable customer demands.

As we showed in this chapter, improvement can be driven by the customer, by the supplier without customer involvement, or by a coordinated customer-supplier team effort. While we again emphasize the latter as the most logical, practical approach to improvement, its absence will not prevent improvement. This is in line with the principle of variation management, which is based on the premise that even externally driven sources of variation can be managed through internally driven actions.

We have also shown that the aerospace industry is early in its understanding of the ability it holds to produce dramatic improvement in such a pacing area. While a number of facilities implemented supplier improvement programs, many of these efforts were based on imposing stringent requirements on their vendors. Quality and delivery performance became closely measured, and cost reduction goals were required as a means to weed out problem suppliers. In most cases, these strict measures were put in place without visible effort to understand the mitigating constraints that could affect a vendor's ability to perform.

As a result, the large supplier reductions that have occurred across the industry during recent years may not have achieved their desired effects. Instead of eliminating those that were limited by real factors—such as insufficient technology, skills, or management capabilities—many may have been forced out for entirely different reasons. Meanwhile, those in stronger financial situations may simply have been better able to endure a short-term squeeze until competitors could be forced out of the business base.

Finally, it must be noted that only a fraction of the overall savings potential that has been identified by this industry's successes is a direct result of supplier improvement. A much broader application of these principles that have already been demonstrated in small pockets could unlock wide-sweeping savings. Thus, the potential for a much larger degree of savings is clear. Those that move decisively in this direction stand to build themselves a substantial competitive edge.

CHAPTER 8

The Role of Product Design

The importance of adopting improved design approaches as a means of controlling product costs has received widespread attention over recent years. It has grown into almost a move-ment as a result of its role in such visible successes as the Japanese automotive industry's dramatic rise to prosperity in the 1980s, and the similar results more recently touted by other industries across the United States.[1] During our own study, we found no shortage of stories extolling gains made through the use of this approach. Dramatic reductions in product development time and engineering change order ac-tivity, and an overall improvement in the fit-up of compo-nents were cited by many that we visited. Only, when we attempted to quantify the specific cost savings to which these have led, we were often disappointed. While it intuitively seems that broader benefits should exist, we found that they are not consistently visible.

This was especially a concern because of the degree of im-portance that these manufacturers place in this area. Many have come to depend on tools and practices aimed at enhancing the manufacturability of their products as the core of their improvement approach. Some appear to have even gone so far as to depend almost solely on these for future savings. Yet our study provides a strong indication that improvements to the process of designing these products may not be the silver bullet to success that some have come to believe.

Should these observations be taken as an indication that there is no real value in pursuing design improvement initiatives? Absolutely not. They should merely be taken as a warning that the savings from implementing improved product development practices may not be sufficient to break the escalation of manufacturing costs. However, they do provide strong reinforcement of the message of this book, that no one of the enablers covered here is sufficient unto itself to facilitate dramatic improvement consistently. Without an integrated approach to improvement such as that identified within this book, the type of substantial cost improvement that must be achieved will likely remain elusive.

Benefits of Product Design Enhancements

In general, the reasons for focusing on this family of design tools and practices appears relatively straightforward: By placing a greater emphasis on one's ability to manufacture a product as early in the design process as possible, greater efficiency can be realized when it is ultimately produced. Clearly, the design of a less complex product, one made up of many fewer individual parts, will reduce the time needed for assembly. Incorporation of more readily available standardized parts can reduce the cost of supplied components as well. By leveraging the knowledge of manufacturing personnel during the design process, valuable production lessons can be rolled into the product configuration. Consequently, fewer lessons will be left to be discovered once production is started. The areas of payback seem to be intuitively evident: There will be fewer start-up pains during the initial stages of production, with a design that is ultimately easier—and cheaper—to put together.

In fact, we have found no shortage of historical accounts supporting the need for this type of emphasis. Deficiencies and oversights during product design have been ascribed as one of the leading causes of disruption to initial production operations. In some cases, the failure of designers to match dimensional tolerance requirements or other important features to the factory's processing capabilities have led to substantial manufacturing problems. Low production yields, poor assembly fit-up, and high acceptance test failures can spur the need for increased inventories, production times, and labor. An inability to provide suppliers with a stable design can lead to problems with configuration control, with a direct impact on their ability to support production schedules. This disruption can artificially inflate the man-hours required to assemble the first units off the production line; this can play havoc on learning-curve projections, casting doubt on estimates for downstream production costs. Ultimately, it can lead to an even more severe consequence: the loss of customer confidence, ultimately affecting their support for the product.

The impact to the cost of production is most pronounced when design deficiencies remain unrecognized until it is too late for them to be corrected. This is especially problematic in the aircraft industry, where the cutoff point for production changes is often reached well before the first unit is manufactured. Because of the need for rapid product introduction, facilities must be built, equipment purchased, and costly production tooling often fabricated even as the finishing touches are made to the final product design. In the case of the airframe itself, the very first articles produced are reserved for full-scale structural tests to protect safety during flight tests. Any changes that are introduced once these test vehicles are assembled can invalidate the results, disrupt schedules, and potentially cripple the program. As a result, even minor changes to configuration or tooling are often not incorporated—even when they offer the potential to streamline production operations—because they are discovered too late to be incorporated without substantial disruption and expense.

There are numerous examples that illustrated to us the disruption caused when designs are "thrown over the wall" too late for those with the experience to recognize potential problem areas to make meaningful changes. With deficiencies discovered only once the first production article is assembled, significant design changes cannot be made. As a result, interim changes (or "Band-Aids") are applied to allow production to continue so that critical delivery schedules and test milestones can be met. Unfortunately, these are often not revisited. Because of this, interim fixes often accumulate to influence manufacturing operations substantially, setting the stage for suboptimal efficiencies for the entire life of the program.

From this discussion, it is apparent why it has become so widely recognized that much of the cost of a product is determined early in the design process. The opportunity to optimize the product configuration for production ease is clearly possible, and can only be accomplished as a part of the basic design process. It cannot be viewed as an add-on, once the design begins to mature, the types of improvement that can dramatically reduce the cost of production are no longer possible. Any inefficiencies that are not minimized as a normal part of the design process will be captured into the product, only to be successively repeated on each unit for the life of the program. The urgency of considering manufacturing implications at the earliest possible time can be easily argued.

In this chapter, we will discuss a number of practices that can contribute to production cost savings by facilitating better coordination during the design process. We will also evaluate the potential for reducing inventories, manufacturing cycle times, and production variation through the use of specifically targeted design tools and practices. We will seek to develop an understanding of the relative contributions offered by each of these areas of emphasis, as well as the interaction with the other improvement enablers discussed in previous chapters.

Enhancing Coordination during the Design Process

As was previously noted, a major cost improvement opportunity lies in the ability to lessen the need for changing the basic product design once it has been frozen. The aircraft industry has historically been plagued by large numbers of these changes, often leading to disruptions to the way in which the end product is to be produced. A major reason for this is the sheer complexity of what must be accomplished—the development and production of a modern aircraft constitutes a substantial feat. Each contains a wide variety of complex systems, ranging from advanced airframes, engines, hydraulics, and electronics to avionics systems with a variety of specialized functions. This design requires skills from a myriad of technical specialties, necessitating substantial coordination throughout its development. Considering the enormity of this task, it is a tribute to aircraft companies that these efforts have consistently led to the development and manufacture of such high-quality, highly advanced vehicles.

This task is made even more difficult by the need to perform to aggressive cost and schedule targets during the development phase. Unfortunately, this can lead to a somewhat myopic perspective. The development phase is often viewed as a separate effort unto itself. In order to devote sufficient effort to achieving performance objectives, longer-term goals can easily be sacrificed. The additional time and tools needed to rigorously evaluate the design for producibility features, for instance, must compete with other objectives that may be perceived to have greater significance. Little time is available to coordinate details across the design teams, with even less available to involve the less mainstream disciplines. As a result, problems may be found only when it is too late to take corrective action.

This mind-set can have severe consequences later on when the product has moved to full-scale production. If these pressures prevent a substantial up-front emphasis on the control of life-cycle costs, it is unlikely, as noted earlier, that many of these considerations can be adopted later. However, recent shifts in the environment have substantially increased the importance of life-cycle costs to the customer. As a result, a change in approach is needed.

THE ADVENT OF INTEGRATED PRODUCT TEAMS

At the core of the shift to improved design practices is the use of an integrated product development approach. Adopted by much of this industry, the premise is simple: By better coordinating the design process across all technical disciplines, disruption caused by oversights can be reduced. Thus, the need for design changes will be dramatically cut, resulting in the need for fewer "Band-Aid" fixes. Ultimately, it is expected that this approach will allow the quicker introduction of a more producible, higher-quality product design.

While specific approaches to implementation vary, the

process generally involves the establishment of multidisciplined teams to develop each of the major portions of the product (in the more advanced applications, major suppliers are also included in this process). A key aspect of this approach is the colocation of personnel from a range of departments to form project teams. These teams can be established for each of the key product components, allowing all to be involved in decisions affecting product configuration, functionality, and performance. As a result, trade-offs emphasizing one area over another are made only as a result of a rigorous decision process. Thus, the importance of such areas as production efficiency, reliability, and quality can receive greater emphasis. Because the teams are colocated and can share design information, several tasks that were once conducted sequentially can now be performed in parallel, thereby reducing the time required to complete the design.

The makeup of the team is as important as the approach itself. For example, suppose that someone from the purchasing department is included in the team. That individual's presence during the decision process will likely lead to a greater emphasis on utilizing existing, common parts wherever possible. The benefits of this are widespread: Along with streamlining the job of supplier management, this reduces the number of unique parts contained in the product, and may reduce the complexity of the bill of materials—ultimately leading to reduced variability in the production process. From our discussions in previous chapters, this has the potential to substantially affect manufacturing cycle time, inventory, and cycle time variation.

It is interesting to note that this approach represents a throwback to the methods employed during the early years of aviation. When aircraft were much simpler pieces of machinery, the same engineers who designed these products worked with production personnel to resolve problems as they were discovered on the production floor. Together, they could develop a knowledge base from these lessons learned that would later be rolled into future design efforts. In his memoirs, one Grumman executive reflected on this approach during his early days as a design engineer:

> Part of my time I would be on the factory floor, working out detail problems with shop foreman Frank Baerst and the mechanics who were building the vehicles. This close liaison was part of Grumman culture; engineers and factory people appreciated the importance of each other's problems and worked together for successful results.
> —Peter Viemeister[2]

After the war, coordination with manufacturing organizations evolved to become a function left to a group of liaison engineers, separating designers from direct interface with the production line. While this allowed the project engineers to better focus on their primary task of designing aircraft, an essential communication linkage was broken.[3] While those most directly involved in production became the most knowledgeable on producibility issues, their role in new product design efforts was limited.

For decades it appeared that, despite these failings, this approach remained the best method to design such a complex product as an aircraft. Only now, with the emergence of new experiences and enablers, has the industry been able to again revert to an approach emphasizing up-front coordination with those who must ultimately produce the product.

LEVERAGING TECHNOLOGY-BASED TOOLS

Only recently have technology-based tools become sufficiently advanced to serve as one such enabler. While major advances in design tools have long been pursued within the core design disciplines, this emphasis has recently taken hold in areas related to controlling product costs. Many have begun to find that these tools and practices also hold promise in identifying considerations that can later streamline manufacturing operations. In fact, many of the tools originally intended for different purposes are being adapted to enable these newer goals. As a result, a substantial tool kit has emerged in a very short time frame.

Among the computer-based tools that fall into this category, digital databases are the most noteworthy. Initially developed primarily as a drafting tool, these same databases are now being leveraged to allow for the concurrent evaluation of product configurations by all on the design team. This capability to allow literally hundreds of individuals access to a drawing before it has been released—without disruption to the design effort—has been repeatedly identified as a major step forward in the movement to integrated product development. With this tool, there is no longer any need to cut out anyone from the effort. No longer is there any reason for designs to be "thrown over the wall" to noncore design groups, only for oversights to be discovered too late for correction. In many cases, this capability can be provided by merely an extension of the equipment and information already resident within the facility.

Some facilities have further leveraged this capability to include simulation tools, permitting design teams to digitally check the product's nominal configuration for fit-up problems even before a single part is made. In addition, such functions can be performed as matching design details to existing process capabilities, ensuring that they can be cost-effectively produced within existing facilities. Others have taken this even further, using the system actively to prevent design conflicts. After sets of key characteristics have been identified by all members of the integrated product team, the digital database is adjusted to prevent designers from exceeding manufacturing limits.

Still, we found that the power of these tools is often not fully leveraged. While many of these systems have the capability to predict the impact of tolerance and process variations, this capability has often not been utilized. Instead, many almost single-mindedly focused on the product's nominal configuration, leaving the effects of this variation to be dealt with only when discovered through the physical assembly of the product. In these cases, design teams seemed to be satisfied with the level of improvement possible simply through better coordination of this nominal configuration. We found, however, that without adequate consideration to built-in variations from this nominal condition (or manufacturing tolerances), a number of problems arose.

As would be expected, the greatest need for this consideration was in highly complex design details. We found no shortage of cases where tolerance buildup led to major problems in assembly, especially at complex interfaces. Often, these issues paced production operations because of their direct impact on manufacturing cycle times as well as on production schedules. In order to control these production delays, substantial resources were assigned, usually leading only to "Band-Aid" fixes that offered less than optimal results. Ironically, the facilities that fell into this trap typically had access to the type of tools to prevent this occurrence in the first place (such as Variability Simulation Analysis software similar to that used in the automotive industry).[4] Because attendant delays were often at the workstations where the product was built up to nearly its highest value state, the costs could be substantial.

It is not our intent to state that *only* through the use of these high-tech tools can dramatic improvement of the design process be achieved. In fact, we have found that some have migrated to the other end of the spectrum, utilizing much less sophisticated techniques with exceptional results. Manual checklists and score sheets were sometimes used to ensure that predetermined producibility considerations were reviewed at key points in the design cycle. A range of low-tech tools have been developed for the purpose of coordinating key design details among the members of multidisciplined design teams, ensuring the awareness of critical interfaces and considerations even before the design takes shape. Others even utilize low-tech factory floor simulations to ferret out the most effective factory flows for these products, rolling this knowledge into the design to make it ever more manufacturable.

Regardless of their complexity, we found that, when logically applied, these design tools have consistently led to real improvement. However, in addition to these tools, we found that the adoption of risk reduction practices also greatly enhances overall results.

RISK REDUCTION PRACTICES

Despite the application of enhanced coordination practices and design tools during the development process, design-based production problems can still occur. The use of state-of-the-art materials, equipment, and fabrication techniques to achieve cost and performance goals can have a dramatic impact on production risk. Each unproven process introduces a degree of uncertainty that can impact production. In order to manage these risks, some have found it advantageous to adopt a methodical, building-block validation program as an integral part of the overall design program. Incremental validation of the processing characteristics of new materials, fabrication methods, and advanced assembly concepts enables problems and limitations to be understood sufficiently early to minimize these risks.

By appropriately sequencing this validation program, results can be seen sufficiently early so that adjustments for any lessons learned can be rolled back into the design before the product configuration is finalized. This leads to two important advantages: Critical delays due to design changes at the start of the production program can be avoided, and designers can feel free to take more risks in their designs. This safety net enables innovative approaches to manufacturing to be attempted without the fear that a failure will have widespread program implications. With this approach, it is more likely that innovative fabrication, tooling, and assembly methods will be attempted, potentially leading to large gains in manufacturing efficiencies.

We have seen a tremendous range in the effectiveness of this approach across the aircraft industry. Those that have most benefited from it typically maintained a rigorous, disciplined approach, incrementally building from the validation of materials to their assembly processes. Areas of focus were logically developed based on an evaluation of their own past performance, but balanced with an understanding of their experience with related materials, processes, and approaches. On the other hand, benefits were less apparent when this approach was implemented without a well-organized plan, properly sequenced to allow feedback in sufficient time to affect later iterations of their product design activities.

Some facilities have extended this risk reduction approach into the production process itself. We found some tooling and work processes designed such that products can be incrementally validated as they move down the production line. Portions of the end item acceptance tests are conducted incrementally at the earliest practical stations during assembly, thus dramatically reducing the risk of acceptance test failures at the end of production. This practice controls manufacturing costs in two ways: It affords greater access to correct discrepancies before assembly is completed, substantially reducing the complexity of these corrections; and attendant delays can be found at production points that represent a lower product value state. As was discussed as early as Chapter 1, it is more efficient to have necessary delays occur at this lower value state, since they will result in a lower carrying cost penalty (as depicted in Figure 8.1).

Measurement of Cost Savings

As was mentioned at the start of this chapter, we found that the actual cost savings resulting from these tools and practices have sometimes fallen substantially below expectations. Upon further investigation, we found at least one very good reason for this: The savings expectations themselves are often poorly defined, frequently based only on anecdotal evidence. Often, facilities had not even attempted to project

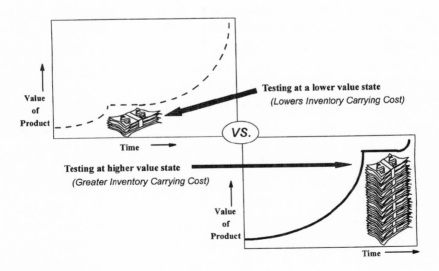

FIGURE 8.1 Advantages of correcting defects at a state of lower product value.

the specific benefits they expected to realize, instead proceeding with plans based merely on a general hope of improvement. Without specific goals, enhancement of performance becomes very difficult to gauge. In the absence of solid data, cost savings cannot be reliably validated.

We had great difficulty finding sufficient evidence to attempt a measurement of success. The following are some of the largest contributors:

- Few new development programs were sufficiently mature to generate data validating production savings.
- Cases where direct measurement is possible are limited. Direct measurement of benefits can be made only in cases where different approaches had been used to develop comparable products.
- Limited data has been collected from which to make before/after comparisons.

Despite these limitations, a few examples were found where the data necessary to directly measure improvement was available. From these, we were able to see something that we had not expected to find: The degree of improvement made as measured against our core metrics—manufacturing cycle times, inventories, and cycle time variation—varied considerably. We recognized that, in order to credibly project the savings potential from applying the tools and practices of this enabler (Emphasis of Manufacturing in Design), we must understand the reasons for this variation.

We found cases of unusually large gains to be the result of both direct and indirect influences. We define the direct influences as those attributable only to an improvement to the design. As an example, a change allowing machining tolerances to be relaxed may allow processing requirements to better match the capabilities of a factory's equipment. Without this change, the factory would continue to be disrupted by processing problems, with parts regularly produced outside of processing limits.

Alternatively, we define indirect influences as those that could have been generated in part or wholly through actions unrelated to these design improvements. In facilities with substantial disruption to their production operations, these indirect influences can be very large.

As an example, consider a factory subjected to a great number of inefficiencies associated with poor inventory and manufacturing control. Any action taken to reduce the number of parts that must be fabricated, tracked, and assembled within this facility reduces the product's exposure to a variety of inefficiencies. By simply reducing the number of items that would have been managed by this system—a system that carries bloated inventories for each item—overall inventories will drop substantially. Thus, while a poorly controlled facility still operates at much higher costs, the introduction of a simpler product design into this facility may give the illusion of a greater payoff.

As was discussed in Chapter 3, while many of the facilities that we visited had developed individual products with simpler designs, they may not yet have successfully implemented such foundation enablers as a solid degree of production and inventory control. As one may expect, enhancements to product designs sometimes yielded exceptionally large improvements. It was evident that indirect influences made some contribution to these results.

Because of inconsistencies in factory environments and the highly variable results they caused, our assessment of the value offered by improved design practices did not result in the absolute quantification of potential savings. However, this in itself is an important finding: It underscores the existence of an interrelationship between design improvements and factory enhancements in controlling the cost of production.

Synergy between Improvements to Design and Factory Operations

In order to demonstrate this interrelationship, we studied the tools and practices used across this industry in current or recent product design efforts. In fact, we found that these tools and practices can be broken into two distinct categories: a larger group containing those intended to improve a factory's basic ability to produce products repeatably, and that subset intended to more directly target the streamlining of factory flow. These categories are illustrated in Figure 8.2.

The difference between these two categories can be highlighted by contrasting the efforts of two facilities as they sought to simplify their product configurations. In one case, a straightforward effort was made to reduce the number of parts that made up the product. In the other, a similar effort

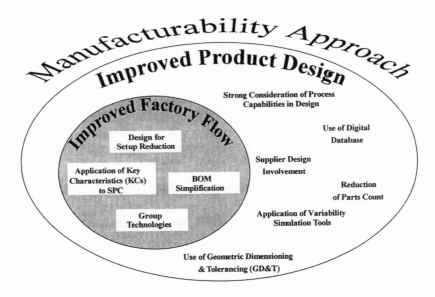

FIGURE 8.2 Types of design initiatives.
Source: The Manufacturing Affordability Development Program, Final Report.

first focused on flattening the bill of materials. While both efforts yielded large parts count reductions the latter led to a greater reduction in the number of factory routings necessary to support its production. Thus, while both improved product design, the latter did so with an emphasis on improving factory flow.

This can be better explained through the use of an example. Consider the construction of a generic aircraft fuselage section. Suppose that its aluminum skins are supported by a complex arrangement of bulkheads, frames longerons, and other structural members. Now suppose that these are each built up from smaller assemblies, many of which are fastened together from formed sheet metal elements. At each level within this very deep bill of materials are vast quantities of fasteners, clips, brackets, and other miscellaneous parts. You can begin to see that the structure is very complex, with a bill of materials consisting of many levels of indenture.

The factory routings necessary to support this complex buildup are similarly complex. A number of routings are likely needed to form the range of sheet metal substructural elements (different operations are likely needed to form curved and angled parts of varying thickness, size, shape, and base materials). They must then move to another shop for heat treatment, and again for such operations as surface priming. These and a variety of other components must then find their way to different factory locations to be fastened together into larger structural elements. This entire sequence begins again for another series of parts necessary to load a tool to yield the next higher assembly. After possibly several more such iterations, the fuselage section is completed.

Now, suppose the producer of this assembly decided to simplify this design by reducing the number of parts used in its construction. With an end goal of simply reducing the parts count, his efforts will seemingly be best rewarded by eliminating those items most commonly used. However, by simply targeting these items—such as the clips, brackets, and fasteners—he may succeed in significantly, reducing the size of the parts list while making only a moderate impact on the number of subassembly buildup step.

Suppose he instead focused on minimizing the number of subassemblies that must be built, thus reducing the number of sequential assembly operations. This objective would tend to drive him to a configuration that more broadly unitizes the structure, possibly shifting from the use of sheet metal in favor of incorporating many of these details into large machined parts—or even single-piece composite structures. By minimizing the number of levels of buildup required, he would also slash the factory complexity necessary to support the fabrication and routing of the many component items.

While our example focuses on the construction of airframe components, the same principle equally applies to such commodities as engines and electronics. In these cases the minimization of levels of assembly buildup may also translate to reductions in sequential test requirements, thus further reducing factory complexity.

With this distinction in mind, we set out to understand the relationship between current production successes and the effects that these may have had on the types of design practices subsequently adopted. In order to search for trends, we compared data as shown in Figure 8.3. On the vertical axis, we have arranged this industry's tools and practices, roughly based on their placement within the categories identified in Figure 8.2. (This is not intended to indicate a relative prioritization within these categories, only general placement within them.) These are compared to the relative success of factory-based initiatives in improving the performance of our three improvement metrics: manufacturing cycle time, inventories, and cycle time variation. As shown in the figure, a definite correlation can be seen from this.

From this chart, we can see that those initiatives intended

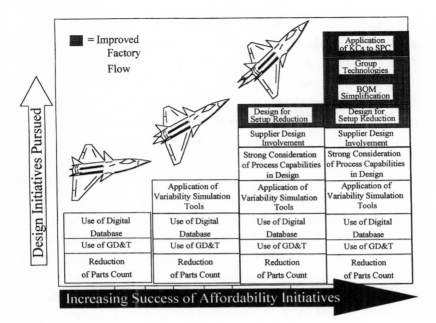

FIGURE 8.3 Effects of factory success on design initiatives pursued.
Source: Based on a similar figure from *The Manufacturing Affordability Development Program,* Final Report.

to improve factory flow, such as application of key characteristics to SPC, group technologies, and bill of materials simplification, are much more likely to be applied within a facility that has already achieved a strong degree of success with initiatives aimed at improving factory efficiency. As was noted earlier, by actively working to simplify their bill of materials, some facilities leveraged the more common practice of parts count reduction to achieve much farther-reaching goals. Through the active consideration of group technologies by design teams, streamlined production flow was sought through the minimization of production routing complexity. This seemed to help optimize the results of a number of production flow enhancement enablers, such as cellularization, automation, and setup reduction initiatives. By giving early, substantial consideration to their overall production flow improvement goals, they appear to have focused their design efforts to influence this outcome directly.

Clearly, these facilities have gained an appreciation for the impact offered by improvement to factory efficiency. By applying their lessons gained from implementing factory improvements, they appear to have a much better understanding of the specific outcomes that they expect from their development program. As was noted in Chapter 3, rather than pushing toward some foggy destination, they now have a focused image of what they are attempting to achieve. While it is too soon to measure the outcome, it seems that this sharper focus will result in a substantial payoff.

We found evidence suggesting that these enhancements can lead to a shift in the production learning curves for affected products. Reducing the early disruption that traditionally exists due to poor coordination during the design process can considerably shorten the time required to produce initial production units. As is notionally depicted in Figure 8.4, this would permit a production learning curve to begin at a much lower point than would otherwise be possible. By factoring in the additional benefits offered by the advanced design practices previously discussed, an even more substantial effect is possible. As was previously demonstrated by Figure 5.5, the strong application of variation management's foundation enablers can result in a shift in the slope of this learning curve. Thus, it seems to be possible for this factory to leap not only to a much lower learning curve, but to one with a steeper slope. While these effects were not possible for us to demonstrate in combination because of inadequate supporting data, it only stands to reason that the effects we observed separately can be combined. Together, they have the potential to produce an even greater cost savings than we have shown.

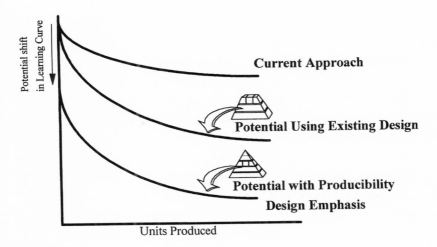

FIGURE 8.4 Improved affordability for new designs.
Source: Based on a similar figure from *The Manufacturing Affordability Development Program,* Final Report.

It is important to note that we use learning curves to illustrate the effects of these improvements only because of their widespread acceptance across this industry. In fact, we have found that it has been this means of displaying improvement potential that has most often gained the attention of our audiences. Perhaps this is partly because of learning curves' long history with this industry; they were first used to predict efficiency improvements that were found to occur for successive aircraft as they rolled off production lines during World War II.[5] Since that time, they have gained widespread acceptance as a means to generally predict improvements to efficiency as production operations mature (resulting from such widely ranging factors as adjustments to factory layout or production control, tooling improvements, and even automation). In theory, this gives a manager an ability to broadly predict improvement trends without the need to thoroughly understand those specific activities that cause these gains.

Yet, our own findings lead us to question the continued suitability of learning curves as a management tool as this industry continues to move forward with lean manufacturing. With individual starting points and slopes so heavily dependent on such variable factors as the producibility of a product and a factory's control of its production and inventories, their construction seems far from precise. For this reason, factories in various states of improvement will likely find that their predictions—which are based on historical trends—no longer hold true; this can result in a wide range of adverse effects. Perhaps this industry will ultimately find that learning curves should be abandoned in favor of a more precise approach.[6]

Contribution of Design Improvements

From our discussions, it can be seen that determining the real contributions possible from design improvements is not as straightforward as some may believe. No simple equation can be used to calculate the effect that design improvements will have on the costs of producing the product. However, we do know that the amount of savings that a facility will see depends largely on the current efficiencies of its production operations, and how it applies these practices while developing its products.

Those facilities that had made substantial progress in implementing first-level enablers appear to be best positioned to reap the greatest savings from new design efforts. While the industry in general appears only to have a vaguely defined focus for these efforts, these facilities appear to have a much better understanding of the specific results that they expect to achieve. As with any effort, a clear vision of what is ultimately sought generally results in a much greater chance of success.

Regardless of the degree of focus, this attention during the design phase has almost always produced some indication that improvement resulted. Improved fit-up on initial production units and reduced rates of scrap and rework were frequently demonstrated. Historical comparisons with similar previous efforts showed large reductions in the overall time required for product development. One of the most striking indicators was the greatly diminished level of design change activity—sometimes to a fraction of the historical values.

As a final note, it appears that others have begun to understand this interrelationship between factory process improvements and product improvement. In fact, the importance of process improvement as a precursor to optimal product improvement efforts has started to be acknowledged. Figure 8.5 can be interpreted not only as the author

THE SECOND WAVE

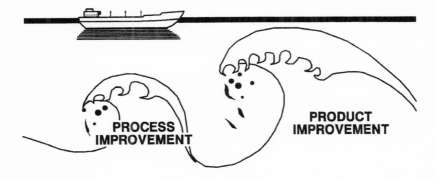

FIGURE 8.5 Relationship between process improvement and product improvement.

Source: Reprinted with the permission of the Free Press, a Division of Simon & Schuster, Inc. from *Reinventing the Factory II: Managing the World Class Factory* by Roy L. Harmon (pp. 154–155). Copyright © 1992 by Roy L. Harmon.

had intended—that process improvements appear to be preceding product design changes—but also as a depiction of a preferred sequence of implementation. By first streamlining factory operations, the true savings potential from design improvements can be better understood, allowing design initiatives to be better targeted.

CHAPTER 9

The Culture of Manufacturing

To this point, our discussion of the path to manufacturing cost reduction has been focused on the mechanics of the job, or the tools, processes, and practices that enable this type of reform. We must now acknowledge that this alone does not constitute a complete picture. Many organizations have invested great energies, resources, and attention, yet have still failed to reap the bulk of the benefit that these approaches have been shown to offer. Those most successful in this endeavor have recognized that an additional consideration is necessary, one that cuts to the core of their entire improvement effort. These organizations have recognized the *criticality of their workforce*, that the efforts of their people are integral to the ability of the organization to change in any meaningful way.

Upon reading our initial study, many industry representatives expressed great satisfaction that we acknowledged the importance and challenges of bringing the workforce through a transition to enable major change. Persuading the workforce to actually accept and *use* the new practices and tools is not a trivial task. After all, this type of transition calls for workers to move out of their comfort zone; it requires them to let go of approaches with which they have long been familiar, and to adopt others with which they are not. They are asked to begin interfacing with new tools, coordinate with others differently, and generally alter the way they spend their time. Often, they must build confidence in new methods and practices based on nothing more than assurances from their management. They must abandon the systems that have allowed them to perform successfully against the goals to which they are measured, and instead move toward approaches that they may perceive to be no different from many others that have in the past fallen by the wayside.

The ability to overcome this skepticism can directly drive the success or failure of a new approach. Typically, one does not strive to utilize something that one does not first accept as useful. Yet, we found that only if the workforce buys in and wholeheartedly uses a new approach will it work effectively. This can create somewhat of a paradox for managers: They must try to get workers to honestly use something they don't yet trust.

One must bear in mind that it is these same workers who perform the functions necessary to keep the facility functioning on a daily basis. The manner in which they complete their tasks is often the driving force behind the success of practices, tools, or systems that enable each of the elements of our improvement framework (Figure 3.6). Thus, their support is crucial to any factory-wide transition. If workers choose to neglect a management tool, such as an MRP II system, the validity of the information on which it is based can rapidly deteriorate, eventually leading the tool to become virtually useless. Without a firm commitment to use it, the system is never even given the opportunity to succeed. Apparently oblivious to this limitation, some facilities continue to revise and reimplement these systems in the same manner, only to see them fail over and over again.

Failing to address the true causes of their initial failures can lead to wasted effort, suboptimal solutions, and unnecessary disruption to operations. For instance, in their efforts to reimplement MRP II, some facilities limit themselves to simply updating the system hardware and software. As discussed in Chapter 5, the real driving force behind this type of production and inventory control tool is often the interface with the workforce. When this is not taken into account, we have seen these systems updated at the cost of millions of dollars and countless hours of disruption, only to result in the same problems that previously rendered them ineffective.

In these modern times, it is easy to become fixated on technology as the ultimate enabler to improvement. Yet, as we found with the application of high automation, sophisticated management tools can come to represent a greater solution than is needed. If a workforce limitation truly represents the primary barrier to improvement, it must be treated like any other factory bottleneck and addressed accordingly. With this limitation corrected, it may be found that a less sophisticated system can adequately service the needs of the facility.

In the aerospace industry, we have found the workforce to be both innovative and proud. Much can be accomplished when these strong characteristics are leveraged in a way that supports ongoing factory improvement. Far too often, this energy is instead directed toward overcoming the deficiencies of institutionalized systems, aimed at inventing work-arounds to preserve schedule and product quality despite the existence of numerous constraints. While inefficient and resource-intensive, these makeshift procedures often manage to achieve the basic goals of the facility. Thus, they gain wide acceptance, becoming deeply ingrained and in time actually replacing the formal management systems.

Rather than expending creative energies to overcome daily problems, a few facilities have been successful in instead focusing this innovativeness toward furthering the capabilities of the factory. Their immediate proximity to the manufacturing operations makes these workers immensely valuable in understanding factory limitations and failures. Lessons can build upon lessons, furthering the knowledge base and refining already solid procedures. If workers are reaching for a real, measurable goal rather than performing perpetual exercises to maintain the status quo, their pride and creative talents can be better stimulated, resulting in a much more motivated workforce.

This outcome is most prevalent where a real effort has been made to ensure the workforce has been properly prepared to receive its new role. Some have found that clear communication on impending changes can go a long way toward attaining this objective. However, we found that this in itself is often not enough; with a culture that has become firmly entrenched over decades of application, a true commitment to move in a new direction is needed. Only as a result of a well-planned, phased approach did we see consistent success in enabling workers to step out of their longstanding roles.

Even more importantly, the workforce must become convinced of *the need to change*. We found that organizations successful in making cultural transitions have consistently demonstrated change to be in the best interest of all within the organization—both the company as well as the workforce. A sense of teamwork is developed, where all see themselves working toward a common goal: the improved competitiveness of the facility. In these cases, a general sense of urgency is typically evident. Often, a crisis of some sort has served as a wake-up call. The loss of market share, the need to shut down a facility, even the perception of increased competition can be used as a rallying point. We found that by sharing with the workforce these reasons on which the direction of the organization is based a greater cooperative environment can result.

Even with this common interest, it is important for all to have a clear understanding of the value of each specific change. Because the workforce is typically composed of intelligent, innovative individuals, it is only reasonable to expect that some explanation must be provided. We found that the better a facility's management can communicate the anticipated results, the more compelling for workers is the reason to change. It seems that only if this is done can individuals internalize and rationalize why they must put themselves through this uncomfortable process.

Our research revealed numerous writings on the subject of managing change within the workforce. Until our own study of this phenomenon, we were skeptical of the true value of these "fluffier" aspects of achieving manufacturing cost control. In fact, only after exploring one particular facility did we begin to understand their importance. After repeated exposure to managers and workers who enthusiastically extolled the virtues of a people-oriented approach, we began to build an understanding of its relationship to their successes. When we reviewed the chronology of facility-wide transformation, not merely in terms of the application of tools and strategies, but in more of a holistic sense, the need for this "people" aspect finally became clear.

One primary reason that this aspect of improvement was so difficult for us to see had to do with its intangible nature. We could not come up with the means to directly measure its relationship to factory improvement. Our metrics (cycle time, inventories, and cycle time variation) did not help us here; we found no clear linkage. Instead, these "people issues" act to facilitate the implementation of other approaches. On their own, they offer no additional increment of savings; however, their absence can limit those savings that are otherwise possible through the application of the improvement framework already discussed in this book.

We have found that each facility exhibits a culture, one that sets the pace for either reform or status quo. Only with the right type of culture in place have we found major reforms that have been successfully implemented. To this point, we have only identified some basic issues that seem to be related to a facility's ability to change this culture. The remainder of the chapter is intended to explore these issues in much greater detail.

Beyond the Cultural Myth

During the 1980s, the U.S. automobile industry began to feel the pinch of an increasingly competitive Japanese automobile industry. This was a far cry from its comparatively strong position at the end of the 1960s, when U.S. industry leaders dismissed the Japanese market penetration as inconsequential.[1] But as the big three and the rest of the country saw the Japanese rise from obscurity to a position that clearly threatened the domestic producers' market share, people began to search for answers. A number of studies sought to determine how the Japanese had managed to improve their efficiency to the point were they could present such a threat.[2] The rigorous adherence to procedures, the painstaking use of process measurement tools, and the dependence on team efforts were all cited as contributors to Japan's successes. In short, much of their progress was attributed directly to the effective use of their workforce.[3]

Some published accounts praised these techniques, while others attributed the Japanese success to more palatable and easily explained causes such as postwar modernization aided by the United States, lower wage rates, and market-penetrating strategies that dumped cheap products on the global marketplace.[4] Others argued that the culture that drove Japanese society itself formed the basis of the phenomenon seen within their factories.[5] After all, this culture was far more rigid, with a premium placed on conformance to society's norms. Stereotypes of Japanese society as a picture of submissive uniformity fueled this growing impression. Clearly, many thought, with our own strengths instead grounded in individuality, the same techniques proven so successful in Japan could not be transferred to an environment so different as our own.

Only much later were these perceptions broken. At first, several early failed attempts to institute these techniques in U.S. factories seemed to underscore the existing belief.[6] This mind-set suffered a major fissure when Japanese companies came to the United States and instituted these same production techniques within American factories—and with American workers. Many, successful in precipitating transformations similar to those seen in Japan, received public accolades for dramatically improving productivity. Today, the belief of the early 1980s—that only the unique nature of the Japanese culture can lead to this result—has been all but dispelled.[7] Even the big three now employ some of these very same techniques.[8]

As we noted earlier, our own study of the aircraft industry turned up examples where similar changes to organizational culture helped enable dramatic, facility-wide improvement. We also noted that these are not widespread. So why is it that some factories can instill and sustain this transformation where other factories have tried and failed? How does a workforce begin the necessary transformation?

After looking across this industry at those companies that have succeeded in making sweeping, permanent improvements across their facilities, we saw a distinct pattern emerge. We found that those companies that had successfully changed the culture of their workforce had three basic factors in common:

- They exhibited great workforce *discipline* in using formal tools and practices.
- A strong *commitment* from the top was clearly visible.
- The workforce increasingly gained *trust* in the direction of the organization.

When all three of these elements are in place, enthusiasm for continued change is clearly visible. Conversely, facilities attempting to implement new initiatives without these tend to demonstrate only localized improvements, with their implementation programs much more susceptible to suffering outright failure.

Developing Workforce Discipline

As was already discussed, we found that many organizations can evolve over time into a culture of work-arounds, abandoning formal operating systems and procedures in favor of informal methods developed simply to get the job done. The laxity that this alternate culture seems to embrace is possibly the single greatest hurdle to implementing any formal factory change. As was discussed in Chapter 5, work-arounds can exacerbate problems in such critical areas as inventory control, data integrity, and production scheduling. Clearly, a major effort to achieving worker discipline is imperative.

As was noted earlier, attaining this discipline can be closely tied to the workforce's degree of confidence in the formal tools and systems they are asked to use. How can individuals be expected to blindly follow rules and procedures if they do not believe in them? This is especially applicable to a situation where a new approach is put in place, one with which the workforce has no prior experience.

We found a couple of organizations that had virtually eliminated this culture of work-arounds, establishing high levels of discipline even where it did not before exist. They did this by ensuring that three factors were considered in any of the new practices that they adopted: a rigorous implementation approach; clear, measurable goals against which progress is closely tracked and reported; and early, visible successes.

THE IMPORTANCE OF A RIGOROUS IMPLEMENTATION PLAN

The implementation of many of the approaches that have been discussed in this book depends on the efforts and expertise of individuals across the organization, as well as on the timely completion of a complex succession of individual tasks. As a result, the coordination of resources and schedules can become a significant management exercise. As is the case when managing any complex project, rigorous planning will dramatically increase the chances of success, while at the same time it will minimize the factory's exposure to disruption.

We have seen clearly demonstrated examples of the benefits of adopting an integrated implementation plan prior to attempting such an undertaking. Through the use of tried-and-true methods of project management, individual tasks can be rationally sequenced and prioritized, with milestones and decision points realistically selected. As a result of this rigorous methodology, the necessary resources for low-risk implementation can be identified and assigned from the start.

This rigorous approach can also go a long way in creating an understanding across the workforce that workarounds will no longer be tolerated, that the approach being implemented is now *the* way of doing business. When it is shown that workers will now be accountable to a new set of performance standards, one that can be achieved only through the use of this *primary* system, they will likely feel additional motivation to work through—instead of around—the system.

We found this sort of rigor to be essential. Without it, chances of success are dramatically reduced. We observed cases where substantial problems have been encountered, sometimes turning the workforce off to otherwise positive innovations. This result seems to be particularly harmful; as we have noted, this can disable any future attempts to implement similar improvements. In one case, we found that a facility actually had to rename a major initiative just to resell it to the workforce.[9] From this we can see that it can be a major mistake for an organization to err on the side of action. It is much more prudent to fully understand the steps that must be accomplished before embarking on any new initiative.

CLEARLY MEASURABLE GOALS

In order to ensure the success of any complex program, one must have in place the means to measure progress. Measurable goals established at the outset provide a baseline from which to gauge success or failure at each phase. We found that it is important to have two types of objectives: overall and incremental. The overall objectives set a target for what must be reached at the end of the project, such as cycle time and inventory reductions. Incremental objectives can be used to track progress at each phase of implementation. For example, they can track inventory knowledge, identifying when inventory accuracies become sufficient to permit low-risk implementation of follow-on enablers. Rigorously tracking implementation using these types of strong, definitive metrics enables any surprises that might otherwise derail the project to be identified in time to allow for recovery.

We found that these goals must be sufficiently specific and clear to allow true progress to be gauged. When they are made available for all in the facility to see, these metrics can provide essential feedback to reinforce the benefits of the new approach. With this progress made visible, the workforce will have the opportunity to become encouraged due to the progress made by their efforts.

EARLY EVIDENCE OF SUCCESS

For complex undertakings, the demonstration of early, decisive success appears to be crucial to gaining workforce buy-in, and ultimately to establishing a disciplined use of the system. Even small gains against incremental milestones can be a crucial means to securing broad-based support among the workforce, underscoring forward progress and management focus. The simple act of tracking and reporting progress against metrics sends a strong message—one of constancy of purpose. By overcoming doubt that this approach will be made to work, management ensures that the workforce will no longer have a reason to develop work-arounds, and the new system will be given the chance to succeed.

A well-sequenced implementation approach can go a long way to allowing an organization to leverage this effect. Careful selection of milestones and decision points to demonstrate increments of improvement means that the workforce can be regularly appraised of progress as well as its impact on the organization. A serious effort to train the workforce can pay major dividends here; in addition to providing value by demonstrating a strong commitment to the people, it will enable them to better understand the meaning of incremental successes. As a result, the organization can progressively gain enthusiasm for the new approach, even before it has demonstrated its full value to the facility.

This iterative approach to gaining workforce buy-in is illustrated in Figure 9.1. If an organization sets out to gain some degree of acceptance before the system is put in place, it will be more likely to achieve some degree of initial success. This proof of the system subsequently leads to greater worker buy-in, further improving the results. The better the results, the greater becomes worker confidence—ultimately leading to a disciplined use of the new system.

One facility that we visited related a story to us that demonstrated precisely this effect. Because of a history of shortages of shop consumable items, workers became accustomed to hoarding them whenever they became available. As discussed in Chapter 5, this can make it very difficult for

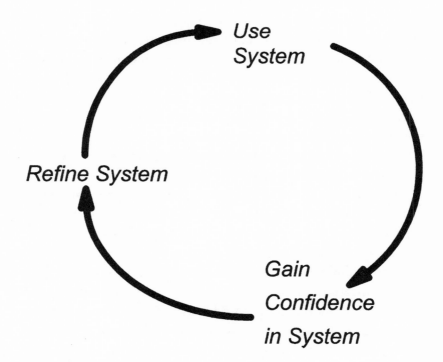

FIGURE 9.1 The culture change cycle.
Source: The Manufacturing Affordability Development Program, Final Report, page 46.

formal production control systems to forecast actual production needs. Major, unforecasted swings in demand make it difficult to keep bins full, leading to regular stock-outs that disrupt production operations. We have seen that this can cause a factory to depend on large safety stocks and teams of expediters to maintain some semblance of order on the production line. Such a seemingly straightforward activity as supplying consumable items to support production operations can erode into a frantic, resource-intensive operation.

In order to reduce the cost of its shop consumables, the facility implemented a program to improve control of its inventories dramatically. Recognizing the need to adopt a solution consistent with its long-term improvement goals, it did not simply resort to the quick fix, one based on the use of strict physical control of inventories. Instead, it set up a rigorous system, carefully crafted to work with—rather than despite—the workforce. By adopting an implementation approach containing each of the elements that we have discussed, the facility quickly gained workforce buy-in, leading to strong discipline in applying the intended system.

As the material control system continually improved the availability of consumable items, the workforce began to abandon the practice of hoarding materials to maintain their own hidden safety stocks. At a point when it was determined that the cultural shift had substantially matured, the organization held an "amnesty day," encouraging the surrender of these private stockpiles in favor of reliance on the inventory system. Because of their newfound belief in the system, workers turned over huge quantities of the items—a result that would likely not have been seen only months earlier. Clearly, this approach to gaining worker trust and discipline had succeeded in transforming the factory culture.

Commitment from the Top

Gaining this type of workforce buy-in can be exceedingly difficult without the understanding that *this* time is different—that the leadership really is committed to change. We have seen numerous cases where management has given its support to an improvement project, only to subsequently lose interest and turn its attention to other matters. This change in focus soon becomes apparent across the factory, and can lead to a quiet demise for even the best initiative. In one instance, this effect undermined a long-term effort to institute an SPC system. Once top management stopped asking to see the SPC charts, they vanished from the factory floor. Without this attention, the resources that had been used to generate them were reassigned to other priorities. As a result, many to whom we spoke have come to believe that, without true commitment from the highest organizational levels, any effort aimed at implementing an improvement initiative is simply a waste of time and energy.

For this reason, we concluded that sustained top-level support plays a crucial role in a facility's ability to succeed in any major improvement effort. During the formative stages, management can set the stage for broad-based results by driving goals and priorities that transcend organizational boundaries and missions. During implementation, managers are in a position to reprioritize conflicting activities, allowing necessary resources to be brought to bear. By simply demonstrating their commitment to its success, they encourage others throughout the organization to lend support, thus minimizing the chance that these undertakings will be deemphasized.

By taking this leadership role, a facility's managers will not only facilitate the introduction of new methods, but also serve as the champions to improvement. By showing this willingness to be an integral part of the process, they can demonstrate their personal stake and commitment, and thus help to forge a path to organizational trust. When interviewing workers in these facilities, we came across instances where a newfound respect developed when workers became convinced that the leadership was clearly behind the effort to improve the facility's conditions.

Gaining the Trust of the Workforce

As has been repeatedly noted throughout this chapter, trust is essential to an organization's ability to rigorously implement new approaches. We found that there are several dimensions to this trust. On one level, the workforce must come to believe that they will truly be provided the means to meet the goals to which they are measured. They must become convinced that, while their efforts may not be immediately visible in terms of factory output, the value of their work will be recognized. With this understanding, they must come to trust that they will not be punished for temporary dips in performance as the factory adjusts to the new system. On a different level, these same people must buy in to the concept that *all* will reap the benefits of the factory's newfound efficiency— that the underlying objective of these changes is *not* the elimination of jobs.

During our study, we found only a few facilities that had been successful in achieving this high degree of trust, especially at the latter level. After all, wouldn't a facility need fewer laborers if new tools and practices proved to be successful in enhancing overall efficiency? We found that measures taken to address these concerns before implementation of changes could lead to a dramatically improved outlook.

A primary means to accomplish this was through the setting up of an extensive training program. Instead of focusing on the further development of the white-collar workforce, these programs often targeted hourly workers, offering a broad range of training opportunities addressing both general and specific areas of study. In some cases, workers were further accommodated with on-site learning centers made available to them both during and after normal work hours.

Why would a facility make such an investment in its hourly laborers? From our discussions with those in management positions, we found that they had truly come to embrace the value of their workforce. In their eyes, if the workforce is to be viewed as an essential enabler to attaining greater factory productivity, substantial attention is justified. If prepared for the factory transition in which they must play a key role, this group of people will be much better equipped and motivated for the challenge. This focus has been shown to have another major payoff: It can go a long way to generating labor-management trust. After all, why would time and money be set aside for their development unless management truly viewed these people as important, long-term members of the team?

The value of this approach is further supported by our findings of Chapter 1. Since the cost of touch labor makes up only a fraction of the cost of production, an emphasis on reducing the size of the workforce offers comparatively low savings. By instead enabling these same people to better address those larger factors driving costs, greater dividends should be possible. This book has identified many of these factors, and has shown the value of a well-trained, flexible workforce in implementing them.

Our conclusion is further reinforced by the gains made by facilities that we found to have adopted this mind-set. They have often been able to implement many of the enablers forming our framework to improvement, allowing them to reap real rewards, often with minimal disruption along the way. Rather than resigning themselves to layoffs and reductions in business, they have seen these improvements in factory productivity allow an expansion of business base. Reduced costs have made it cost-effective to draw work back from suppliers, and in some cases, have generated new customer orders.

Adopting a New Factory Mind-Set

If a mind-set is developed where the workforce and management work hand in hand to achieve this type of common goal, manufacturing costs and ultimately factory competitiveness will likely be dramatically improved. Ultimately, the goals of both are common: to expand the growth, profitability, and long-term security of the organization. Without a cooperative culture, management and labor will have great difficulty in reaching agreements that can enable dramatic improvement in productivity. Those that have found a way to develop this understanding have made great strides in making the best use of the principles and practices of this book.

Some facilities that we visited had made tremendous headway toward this goal. Once, only machinists could operate milling machines, only material handlers could move parts around the factory, and only electricians could touch wiring. By willingly restructuring to perform multiple tasks, facilities have allowed the broader use of the talents of their workforce. Quality inspectors became material handlers when it was necessary, and machinists ran more than one machine, even performing some of their maintenance. By understanding their common need for these changes—the preservation or expansion of the business base—all can feel more comfortable working together to reach this outcome.

It is interesting to note that we found no correlation between the existence or absence of trade unions and the performance of the factories against our metrics. In fact, we ran across a couple of instances where the union brought some of the best facility improvement ideas to the table. This signals that an emphasis in the factors that enable a shift in the cultural behavior of a factory can foster a positive relationship. By concurrently focusing on workforce discipline, organizational commitment, and trust, new approaches can be given the chance to take hold even as the workforce and management come to understand that they are ultimately working toward common goals. In fact, it is sometimes through this process of improvement that all can come to realize that the objective of the facility's management and that of the workforce are not as far apart as once thought.

Still, many organizations have yet to make substantial progress in this area. Perhaps one reason for this is that many do not yet understand the importance of a positive workforce culture in major improvement efforts. In many of those that we visited, efforts in this area were treated almost as an afterthought, as add-ons to programs primarily focused on the more tangible aspects of these efforts. Perhaps our observations will serve to highlight a need that many, deep down, are already aware must be addressed.

A major problem is that progress is difficult to measure. Worker value has been traditionally judged based on touch labor productivity, yet we have discussed in this chapter that this represents only one aspect of what the workforce can offer. Since no real metrics were found to measure directly such intangible characteristics as workforce discipline, these can be gauged only through the use of indirect comparisons. By further evaluating the results of other, more measurable efforts that have been shown to be closely tied to workforce discipline, for example, some means of assessment may be possible, thus mitigating a portion of this challenge.

We recognize that change in organizational culture can represent one of the greatest management challenges. Perhaps the findings provided within this chapter can offer some degree of help. By generating an awareness of limitations in achieving improvement without a concurrent effort to change culture, it is hoped that managing culture change will be more frequently addressed. If they are aware that this change is possible—that it has been demonstrated within the aircraft industry—perhaps others in this and related industries will see the need to consider similar efforts.

CHAPTER 10

Changing the Paradigm

As we discussed in Chapter 2, Henry Ford's approach to automobile manufacturing has had a tremendous impact on industries across the world. His advancement of mass production demonstrated critical practices that would prove to change forever the manner in which products are produced across a range of industries. His would serve as the springboard for further innovation. Even Taiichi Ohno, the architect of the prestigious Toyota production system, credits Ford with inspiring his own efforts.[1] Yet, despite this widespread recognition, Ford's methods did not initially receive a favorable reception. In fact, certain aspects of his approach led critics to go so far as to accuse him of "economic blunders."[2] It was his ability to ignore the preconceptions of the day in favor of an unbiased viewpoint that enabled him to launch his company to a new level of profitability.

Much of Ford's genius was in his pragmatic view of his business. While others focused on controlling costs as they had done prior to the age of mass production—often by squeezing the cost of direct labor—Henry Ford insisted on a different approach. He demonstrated the ability to focus on those areas that could make the greatest impact on profitability. Much as we presented in Chapter 1, he recognized that the answer did not lie in the minimization of direct labor costs; instead, he sought to maximize the productivity of his overall operations.

During a 1914 meeting with his board of directors, Ford listed on a blackboard his plant's overall costs, as broken down by labor, matériel, and overhead. He then proceeded to demonstrate the relative impact of each of these on the company's profits. Much as we have discussed, he showed that the control of direct labor rates received an inordinate amount of attention, since even large increases made only a relatively minor impact on the bottom line.[3] In fact, by making even seemingly major concessions to the workforce, he illustrated that their increased effectiveness would likely drive a better utilization of the facilities, thus substantially increasing factory efficiency. Labor had grown unreliable, with constant turnover draining essential skills and preventing job familiarity.[4] With improved pay and reduced hours, workers would stay in place longer, perform better, and enable the factory to gain efficiency over time.[5] His board, however, had a difficult time seeing beyond their own preconceptions; although Ford presented the benefits of his approach in black and white, it remained difficult for them to swallow.[6]

Ultimately, Henry Ford won the day, nearly doubling the workers' salaries to create the now-famous $5 workday. In retrospect, his approach proved to be correct; by increasing pay and reducing the standard shift to eight hours, he was able to gain a much more capable, stable workforce that conceded to man the factory for three shifts.[7] Factory throughput and productivity skyrocketed. He could now turn out cars nonstop around the clock, with his workers operating at peak performance as a result of their new, shortened workday.[8] Thus, as a result of his ability to see beyond existing paradigms, he drove his company to a new level of profitability.[9]

Much as was the case with Ford's implementation of mass production, a measure of counterintuitive thinking also led Mr. Ohno to develop the Toyota production approach.[10] Yet, even Mr. Ohno was met with a defiant reception when he first tried to implement his system:

> In the beginning, everyone resisted kanban because it seemed to contradict conventional wisdom. . . . When I was—rather forcefully—urging foremen in the production plant to understand kanban, my boss received a number of complaints. They voiced the feeling that this fellow Ohno was doing something utterly ridiculous and should be stopped.[11]

In retrospect, these men would each, in turn, come to transform one of the most significant industries of the twentieth century. Ford's methods quickly led to cutting the price of the Model T almost in half, from $950 in 1909 to $490 in 1913, making it widely affordable and launching the growth of the U.S. automotive industry.[12] Ohno's approach would ultimately lead Toyota to become an international force, with a number of his controversial practices now themselves representing conventional wisdom. Yet, without the ability of each of their companies to look beyond the conventional wisdom of the day and embrace their methods, none of this would have been possible. Similarly, with the principle outlined in this book, aircraft manufacturers must again face down conventional wisdom to transform their industry.

The Paradigms of Today

While the underlying framework of variation management is readily understandable and may be viewed as almost intuitive in nature, this cannot be said for many of the specifics of its application. We have found that key elements of a number of the enablers discussed in the previous chapters defy the conventional wisdom of today, and thus require great forethought and discipline to implement. Facilities that have attempted solid initiatives have fallen short because of these harmful preconceptions, often pressing forward again just to fall into the same trap.

For this reason, we recognize that without a discussion of these paradigms, this book could not be complete. While we cannot begin to cover them all, we will discuss some of the most prominent that we have come across. It is hoped that this will serve to develop a mind-set that will help to rout out similar preconceptions so that the framework set forth in preceding chapters can be effectively and consistently implemented.

INVENTORY REDUCTION PRACTICES

Let us begin by examining the practices commonly used in inventory reduction. The reduction of factory inventory has become one of the leading areas of emphasis for reducing costs. In fact, true efficiency is now sought through the virtual elimination of inventory in all of its forms. Batch production and reorder-point policies are now seen as obsolete tools; one-piece production flow and Just-in-Time (JIT) supply are seen as necessary replacements for these respective approaches.

While minimizing inventories clearly should be a primary objective of a lean factory, we have found that today's paradigm is to target this as an immediate and single focus.

Instead, our discussion in Chapter 5 emphasizes achieving inventory reduction as a primary *outcome* of adopting a disciplined process, rather than through its direct attack.

We have found that only once the most prominent sources of variation are mitigated can a long-term reduction in factory inventories be achieved. Conversely, an immediate attack on high factory inventory has been shown to distract from the correction of its underlying causes, sometimes leading to disastrous results. Thus, inventory reduction should not be viewed as the means to improvement, but as a very visible outcome.

Consider the case of a factory that directly targeted the reduction of stockroom inventory, cutting levels across the board by 20% to 30%. This factory did not follow the orderly approach of Chapter 5; it mandated this action without first developing a solid understanding of its production operations and the reasons for the existence of these excess inventories. This attempt to overcome a complex progression of variation in just a single step ultimately proved to be too much, with stock-outs and work-arounds mounting to the point that operations were nearly crippled. Even a vast team of expediters could no longer control the problem. Only once inventories were increased was order restored.

In contrast, we visited another organization that did not begin its efforts with this emphasis. Instead, it began by implementing the foundation enablers of variation management. As a result it gained a solid understanding of its existing inventories and production flow, ultimately leading to the identification and elimination of excess inventories. However, this approach also highlighted key areas where production operations remained highly unpredictable; in these areas, the organization took the counterintuitive approach of actually *increasing* key inventories. By temporarily cushioning the effect of those sources of variation it had not yet come to understand, out-of-station work and expediting actions were controlled, and production schedules stabilized. With its attention now turned from fire fighting, this organization could now analyze these remaining operations in an orderly manner. As consumption became more and more predictable, these remaining inventories could be progressively reduced.

Using this approach of targeting factory variation as the means of reducing inventory, this facility ultimately cut its inventories by almost 50%—much more and much quicker than we saw in facilities that directly targeted this result. Thus, an approach that some may have seen to be counter to reason was ultimately more successful in achieving this desired effect.

SINGLE-FOCUSED SOLUTIONS

This obsession with inventory reduction as an end unto itself represents only one example of an apparently widespread view that such narrow objectives will translate to large cost savings. We often observed a singular focus on such intermediate metrics as cycle time reduction, equipment utilization, and touch labor efficiency, as well as on various individual enablers discussed in this book. Some believe that these actions simply mimic those of high-profile success stories such as Toyota; our own research has led us to understand that this is no more than a misconception. In fact, in his writings on the subject, the academic force behind the development of Just-in-Time expresses great concern that others might misinterpret these methods precisely as has often been the case:

> If you merely imitate the superficial aspects of the Toyota production system and hastily apply what you see as a system for getting parts just-in-time, you will end up with the opposite of the desired result. This will not only create havoc in your own plant but also cause your suppliers a great deal of stress.
>
> —Dr. Shigeo Shingo[13]

Instead, he and Mr. Ohno describe a path that is not inconsistent with that offered by this book. Much as we have seen in the production of aircraft, Toyota once had difficulty delivering parts to the production line on time. The company's solution began with what was termed "production leveling," or the smoothing of production flow throughout the factory. In theory, setting a rigid assembly schedule would enable the demand for materials and parts to be more precisely planned. Much as we discussed in Chapter 6, by developing more dependable delivery schedules, the assembly shops became better customers to those that supplied materials to them, permitting them to improve their performance. Toyota viewed this step to be truly pivotal; until this "production leveling" had been adequately addressed, the implementation of subsequent improvements—many of which are represented in our higher-level enablers—were not attempted.[14]

This should help to show that the views of Toyota on how to implement its initiatives were quite different from those that represent the current paradigm. The results of this deviation are illustrated by the data displayed in our comparability matrix. From this, we see that many have made strong efforts toward implementing just a limited number of initiatives. The comparability matrix also provides us a clear view of the outcomes that have resulted from this type of approach: A single-focused effort does not result in the degree of savings that would otherwise be possible. Only by implementing these initiatives as the pioneers had originally intended—as part of a structured, thoughtfully sequenced program that progressively attacks the sources of factory variation—can the same levels of achievement be realized.

TECHNOLOGY-BASED SOLUTIONS

Next, consider the current emphasis on adopting technology-based solutions. Many of the facilities that we visited demonstrated a clear preference for improvement approaches that incorporate the latest in technology. Computer simulations, information technologies, and advanced automation were under development across the industry. With all of this effort, we found that there had yet to be the widespread payoff that was sought. Instead, companies continue to expend both valuable time and resources in a struggle to develop the means to apply these technologies.

At the same time, we observed examples of low-technology solutions that seemed to serve similar purposes. Charts taped across the walls of a war room, factory models, and templates of machinery and tooling each serve as inexpensive, rapidly adjustable methods to predict the results of changes to factory layout or work flow. The use of hybrid production control approaches (as discussed in Chapter 5) minimizes the level of automation necessary for factory MRP II systems. Some were even found to be removing their inventory automated storage and retrieval systems (ASRSs) in favor of direct vendor delivery agreements with their suppliers.

These observations led us to the conclusion that, despite the current bias, high-technology approaches do not always constitute the best solution. In fact, we found cases where they are actually harmful to a facility's ability to generate large improvements. Consider a factory that purchases a piece of costly, immovable equipment prior to mastering the first two building blocks of variation management. Now, suppose that as the facility moves toward gaining better production and inventory control, it minimizes those surges in demand that seemingly mandated the purchase of this equipment in the first place. It is possible that by prioritizing these low-technology actions, the need to purchase new equipment could have been eliminated altogether. By now designing production flow around this equipment in order to pay for it, the factory may be permanently locking itself into a suboptimal mode of operation. Thus, this investment in high-technology equipment may have actually prevented further improvement.

A similar argument can be made about the application of information technology. Despite the commonly held objectives of fully leveraging advancements in this area, it can be argued that this type of solution should be sought only where it is absolutely needed because of the high costs associated with its development, setup, and maintenance. As we noted in Chapter 5, the less complex the system is, the fewer problems will be realized where this type of tool is most vulnerable: at the human interface.

Unfortunately, we realize that this extraordinary focus on solutions founded on high technology will not fade. Perhaps this industry has acquired this mind-set because of its roots; after all, every major barrier to this point has been broken as a result of engineering and scientific advancements. This time, however, we have shown the barrier to require quite a different approach: Although technology will continue to play an important role, the first step of the solution must be founded on innovations in the area of variation management. While this approach is not as glamorous and will never be as effective in grabbing headlines, the tremendous outcome it promises should suffice to catch even greater attention.

OUT-OF-SEQUENCE IMPLEMENTATION

Finally, we must note that many continue to implement initiatives in a sequence that is not consistent with that identified in this book. Even some that have seen our initial study and have applauded its conclusions have persisted in applying their old paradigms. We continue to see projects aimed at supposedly demonstrating such initiatives as cellular manufacturing and process automation despite the fact that they have not yet mastered the foundation enablers that should have preceded these efforts.

These demonstrations are particularly disturbing in that they can perpetuate a false belief in suboptimal paths to improvement. Often they are conducted in an environment that is isolated from the primary operations of a plant, making it possible to produce positive examples of gains against a number of popular metrics. Much as was noted in Chapter 4, this narrow focus may not provide a true picture of what has actually resulted. Only when viewed from a factory-wide vantage are the costly inventories or schedule buffers that enable this isolation factored into the results. When this is done, much of the perceived savings will often evaporate. We found that the ability to view these demonstrations in this manner represents the exception rather than the rule.

A common view is that full implementation of any new initiative is risky, and that one's exposure to this risk can be limited to only part of the factory by implementing these initiatives in this isolated manner. Yet, we found that this paradigm is flawed, with many that have adopted this approach sealing their fate from the outset. Consider the discussion in Chapter 5 of precisely this type of partial implementation of MRP II. Because it was not broadly implemented, it initially stumbled—only to succeed once reinstituted as a factory-wide initiative.

What is most ironic about these supposed demonstrations is that the initiatives that they target often have already been proven elsewhere. All that remains to be understood is their relative effectiveness within their own specific environment. From this standpoint, it is difficult to see what benefits are expected, since these demonstration programs seek to isolate these initiatives from their factories. Our hope is that, by disseminating the results of the lessons already learned from those that have truly implemented these initiatives within their factories, the perceived need for further "demonstrations" will be eliminated. With this, the industry can get on with focusing its efforts on implementing the tools and practices of variation management in a manner that will result in a much greater benefit.

CHAPTER 11

A Final Note . . .

In this book we have shown that the aircraft industry has accomplished something quite remarkable. When faced with a serious impediment to its continued advancement, the industry has laid the groundwork to overcome it. Through the combined efforts of individual facilities across the industry, a formalized understanding of an underlying framework for success has emerged, setting the stage for all to overcome this cost barrier. With this, we can now better see why some factories succeed while others seem to wander aimlessly, realizing only minimal gains. We can also see that a range of paradigms must be overcome in order to enable success.

Throughout this book, we have attempted to break the paradigm that the Japanese approach to improvement is based on a group of tools and practices that can be implemented in whatever manner is wished. Instead, we have shown that it is *the method by which these tools and practices are implemented* that distinguishes those that are most successful. As was noted earlier, Toyota did not begin the path to its success by implementing kanban across its factories. It first mastered a series of much less glamorous initiatives that established a foundation from which to work.[1]

In order to gain the most from lean manufacturing, it is critical to develop an understanding of what each of its component enablers is specifically intended to achieve. Each of these represents a powerful means to manage a different source of factory variation. Each must be employed to affect the specific piece of this complex puzzle for which it was designed, in the sequence in which it was intended. Without this, their results become almost hit-or-miss.

The initial focus of our study was on the effects of lean manufacturing tools and practices on the production of a specific end product—an airplane. As a direct result of this focus, we were able to demonstrate the value of applying variation management in implementing lean manufacturing within this industry. We now realize that this was only a first step, one that only begins to leverage the power of this principle.

As our understanding of variation management continued to grow, we began to realize that this principle has broad applicability. Not only can it be used to streamline the production of aircraft, but we have already demonstrated that it can do the same for many other products, ranging from relatively simple to highly complex. It can produce strong results in the production of high-strength airframes, cutting-edge jet engines, sophisticated electronics, and the items that make up these systems. It can streamline the production and delivery of shop consumable items to the production floor. If all of these apply it within the context of their facilities' constraints—as well as their own starting points—they can outline a specific plan for rapid cost reduction.

From these findings, it is not a stretch to say that many others stand to benefit from the application of this same principle. After all, many of the processes and some of this industry's suppliers cross industry lines. When we discussed this with individuals familiar with other industries, they recounted stories supporting this conclusion. In fact, we were often surprised when we heard that these others faced many of the same types of problems that we ourselves have seen. We were met with the familiar question: Given the wide range of lean initiatives available, where should they begin?

From this, we concluded that many other industries likely face the same problems that affect the aircraft industry. For example, supply chain management has been highlighted as a key issue, especially where deep supply chains lead to a great deal of variation in delivery times. Yet, instead of the relatively low-tech approach that can be used to deal with this (as cited in Chapter 7), advanced technology solutions are often chosen. The continued search by companies for other altenatives indicates to us that they are not yet satisfied with their results.

A similar approach can provide the means to slash the life-cycle cost of a product. This is a key area of interest for aircraft industry customers; with the very long life spans of these products, much of the cost of ownership is for spares for in-service support. It is also a key area of difficulty, since demand for spare parts is especially variable, and thus hard to predict. Variation management offers a potential solution here also.

To understand this difficulty, consider the complexity of supporting a production line (as discussed throughout this book). Now, imagine that the customer's demand for the items produced on this line is almost random. This is how many view the demand for spare parts to support the in-service use of a product. Since the need for spare parts largely arises when a part is damaged or wears out, it is very difficult to predict. As a result, forecasting these demands to the fidelity necessary to allow items to be produced just when they are needed by the customer is virtually impossible. Thus, warehouses must be stocked with inventories to support these needs—most likely at a very high cost. Even with this measure, a great deal of effort is needed for "expediting"—just as we found within the factory environment.

Now, suppose the methods of variation management were used (as described in Chapter 7) to arrange for this support. While the demand for individual items may be unpredictable, the demand for similar groups of items will tend to flatten out. Factories that have an improved means to predict demands—not for each individual item, but for a broader group—can preplan their factory flow to accommodate this need. The result is that they can produce a greater percentage of these items just when they are needed. Inventories in warehouses can be slashed, and infrastructure costs can be minimized. Thus, by applying variation management, a seemingly impossible situation can be overcome. The end result is a dramatic drop in life-cycle costs.

We began this book by identifying a concern that many of the problems that prevent this industry from becoming lean remain outside of its own control. It is argued that the demand for its products is deeply cyclical in nature, with dramatic peaks and valleys in customer orders affecting the ability of manufacturing organizations to plan their operations for a smooth, efficient flow. This effect has only been amplified by the end of the Cold War and its implications of permanent reductions in sales to the military sector that have historically helped to dampen commercial demand swings.

In this book, we have intentionally steered clear of any discussion on initiatives aimed at solving this broader issue. Our experience has led us to a sense that the types of comprehensive changes needed to drive any real improvement to this issue will be difficult and will take years to emerge—if they are ever in fact realized. In the interim, however, great strides can still be made. For this reason we have focused on the use of variation management as a means to soften its impact.

Variation management offers a factory the means to improve dramatically its production efficiency—despite the existence if external constraints—by focusing on activities that it *can* control. In the case of the machine shop discussed in Chapter 7, its application of such approaches as group technologies and cellular manufacturing enabled it better to prepare for the types of industry volatility noted. By diversifying its product line, it greatly reduced its exposure to downturns affecting a single industry. Furthermore, its flexible operations provided for a greater ability to quickly ramp up production for unforeseen increases in demand.

By improving their overall performance, facilities that implement variation management may stand to further innoculate themselves from broader industry trends by increasing their own share of the market. Low product costs will likely continue to play into acquisition decisions; we have shown that those who implement variation management have the potential to demonstrate these savings in both production and life-cycle costs. By capturing a greater percentage of the smaller number of new products, they can minimize such substantial problems as overcapacity.

While this ability, in itself, is far from sufficient to insulate a facility from these industry-wide issues, it may greatly soften the blow. It only stands to reason that this ability to better ride out the storm will provide a strong competitive edge.

We will close by stating that the value of lean manufacturing has not been overstated, but the importance of its rigorous implementation may often be underestimated. We have shown that lean manufacturing can offer a great deal of savings across the life cycle of a wide range of products. In order to produce these savings, however, a factory must take on a mind-set that the sources of manufacturing variation are the primary target. When symptoms such as high inventories and cycle times are viewed as the target, much smaller gains, if any, are seen. As with the treatment of a medical patient, actions must be taken that address the root of the problem. When an infection is treated only with aspirin, the primary symptom of pain and fever may be temporarily controlled, but the underlying situation may continue to degrade. Only with a more basic intervention can the patient's long-term prognosis improve.[2]

Appendix

The comparability matrix has served as a key instrument in the establishment of framework to improvement depicted in Figure A.1.[1] It provided the means to visually scan a myriad of data, summarizing it in such a format as to permit comparisons and credible conclusions to be derived. Generic comparison matrices are a commonly used tool for benchmarking analysis, yet the comparability matrix of Figure A.2 appears to be unique in its simultaneous presentation of three characteristics exhibited by each facility in this study. These include the types of tools and practices implemented in each facility; the relative strength with which each of these was applied; and the degree of improvement (as measured as a reduction in cycle time, inventory, or cycle time variation) that was seen during this implementation period.

As was noted in Chapter 3, information gathered during factory visits was recorded on a series of summary sheets. Each of these broke down an individual improvement initiative (such as JIT, MRP II, or supplier reduction, as identified in Figure 3.1) into a number of component pa-

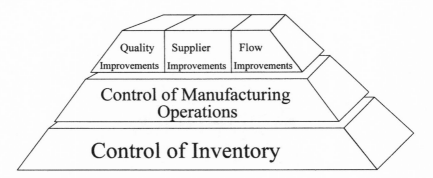

FIGURE A.I The hierarchy of process enablers.

rameters. Each of these parameters was assigned a weighting factor based on its relative significance to the initiative. Finally, each was scored based on the strength of implementation our team observed within the individual facilities, with the parameters summed for an overall score.

We quickly recognized that any meaningful analysis of these numerous and complex data sheets would be difficult. We considered developing a computerized approach, applying complex mathematical algorithms to search for trends. After some thought, we recognized that this would be difficult, and, more importantly, it would likely lead to results based on a range of assumptions that would not be easy to explain.

Instead, we settled on a low-tech approach. We divided the overall scores identified on the individual worksheets for each facility into three categories: the group corresponding to the lowest third in the numerical range was assigned a blank; the middle third was assigned a small x, and the top of the range was denoted by a large X. It was only through the clear visual format offered by this series of symbols used to depict ranges in each of these characteristics that we could readily scrutinize our data in a matrix format.

To complete the picture, we added three columns to the

Facility	Complexity Factor	SPC	Cost of Quality	Metrics	MRP II	JIT	EOQ/Reorder Point	Long-term Suppliers	Supplier Coordination	Flow Time Reduction	Cellular Manufacturing	Automation	BOM Mgmt.	MPS Mgmt.	Workforce Empowerment	Supplier Reduction	Supplier SPC Program	Cycle Time	Inventory (related to sales)	Cycle Time Variability
A	L	x	x		x			x	x				x	x	x	X				
B	H		x					x	x	x			X		x	X	X	↑	↑	
C	M	x	x		x		x	x					x		x	X				
D	L	x	X		X		X	X	X	X	X		X	X	X	X	X	↑	↑	↑
E	H	x		x	x			X	X	X	X		X		x	X	X	↑		
F	L		x	X	x	x	X	x	x	x	x		X		x	X	X	↑		
G	H								X			x				X	X				
H	L	x													X	X		↑		
I	L	x	X		x		x		X	x	X		X		X	X	X	↑		
J	H	X	X	X	X		X	X	X	x			X	X	x	X	X	↑	↑	↑
K	M	x	x	X		x			X	x	x	x	X	X	X	X	x	↑	↑	↑
L	M															X				
M	H			X	X	X	X	X	x		X	X	x	X	x	X	X	↑	↑	↑
N	H	x	X	X			x	X	x	x	x	x	X	X	X	X	X	↑	↑	
O	H	x	X		x		x		X		X	x	X		x	X	X			
P	L	X	x	X	x	X		X	X	X	X		X	X	X	X	X	↑	↑	↑
Q	M	x			x	x	x	x	X				X			X		↑	↑	

Legend: Complexity: H = High, M = Medium, L = Low
Ranking: **X** = Disciplined Approach across the Board,   x = Somewhat Successful,   Blank = Not used/Unsuccessful
Metrics: ↑ = Large Reductions,   ↑ = Moderate Reductions,   Blank = Small/No Reductions

FIGURE A.2   The full comparability matrix.

right of this matrix. These depict the level of factory-wide gains seen as a direct result of the actions depicted by the symbols used in the matrix. Similar symbols were used to visually bound the degree of this savings: No marking depicts very little improvement, if any; a small arrow depicts at least 10% improvement; and a large arrow depicts the largest gains—25% or greater. In fact, this latter category most often represents gains much larger than this (as described in Chapter 3).

Review of this comparability matrix immediately reveals an important trend: Large improvements against the study's three metrics (depicted on the right-hand side of Figure A.2) are seen only where facilities have made *strong* efforts against a range of initiatives (depicted as a large X on the left-hand side of the chart). Conversely, no substantial improvement is

seen against the three improvement metrics for facilities exhibiting weak implementation, *even in implementing a large
number of initiatives*. The data listed for facilities A and C illustrate this quite well; each has implemented more than half
of the initiatives of our study, but their implementation could
almost never be characterized as strong. Both showed almost
no improvement against all of our metrics.

As noted in Chapter 3, dramatic cost improvements were
seen only where we found the strongest level of improvement
against all three metrics (depicted by a large arrow for each of
the three metrics on the right-hand side of Figure A.2). This
was an important finding, since it allowed us to focus our efforts on understanding those activities that correlate to three
large arrows on the right side of the chart (facilities D, J, M,
and P). Similarly, because of our earlier conclusion that improvement against these metrics correlates to strongly implemented groups of initiatives, the relationships between these
and large improvement in all three metrics received the majority of our attention.

In reviewing these relationships for our first enabler, Control of Inventory, we see what can be characterized as a very
strong correlation: *Only* where *all* initiatives in this enabler
have been applied (where appropriate) do we see the type of
results indicative of dramatic cost savings (Figure A.3). It is
for this reason that Control of Inventory forms the foundation of the framework of Figure A.1.

There are a couple of notable considerations in drawing
this conclusion that must be explained. As we indicated in
Chapter 3, factory complexity is an important aspect that
must be taken into account during this analysis. With this in
mind, we must note that facilities D and P are each of low
complexity. For this reason, they have demonstrated strong
performance against all three metrics, yet each did not
strongly implement one of the initiatives within this enabler—MRP II—because MRP II is an initiative that is most

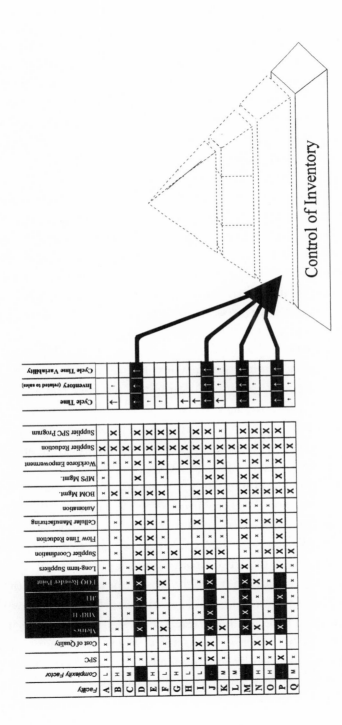

FIGURE A.3 Effect of inventory control on improvement.

227

appropriate for a much more complex facility. It is further explained in Chapter 5 that performance of lower-complexity facilities can often suffer from the use of this system. Instead, the most successful of these have implemented other types of rigorous production and inventory control systems—leading to the types of results seen here. For this reason, a lack of strong implementation of MRP II for these facilities does not represent a weak implementation of this enabler.

It also must also be understood that, as highlighted in Chapter 5, JIT and EOQ/Reorder Point simply represent alternative inventory control approaches. For this reason, we sought to understand if *either* of these had been strongly implemented. With this in mind, we found that each of the facilities that exhibited the strongest improvement against our metrics had demonstrated strong implementation of at least one of these approaches.

The enabler Control of Manufacturing Operations demonstrates a similar trend; when the complexity factor is taken into account (again, affecting how we view the application of MRP II), we again see a perfect correlation. Only when all of the initiatives forming this enabler are applied in a strong, factory-wide manner do we see results that are indicative of large savings (Figure A.4).

For the next enabler, Supplier Improvements, we do not see this same type of correlation. For example, while all of those facilities that participated in this study had substantially reduced their supplier bases, they all did not reap measurable results. Similarly, strong implementation of other initiatives of this enabler (such as supplier SPC and long-term supplier relationships) did not necessarily correlate with strong performance against these metrics. Instead, these initiatives produced strong results only when implemented in conjunction with solid performance on the previous enablers: Control of Inventory and Control of Manufacturing Operations (depicted in Figure A.5).

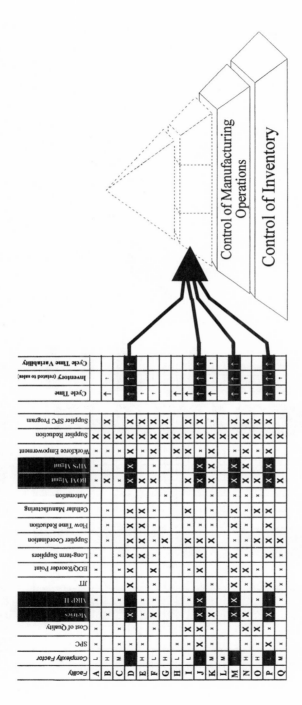

FIGURE A.4 Effect of manufacturing control on improvement.

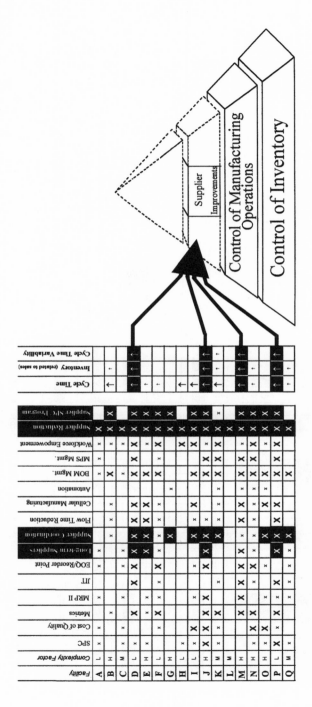

FIGURE A.5 Effect of Supplier Management on improvement.

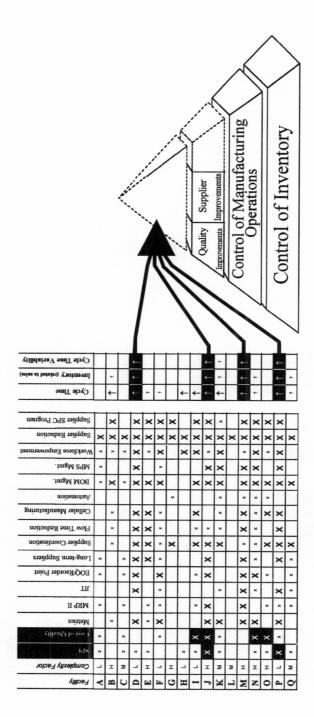

FIGURE A.6 Effect of quality on improvement.

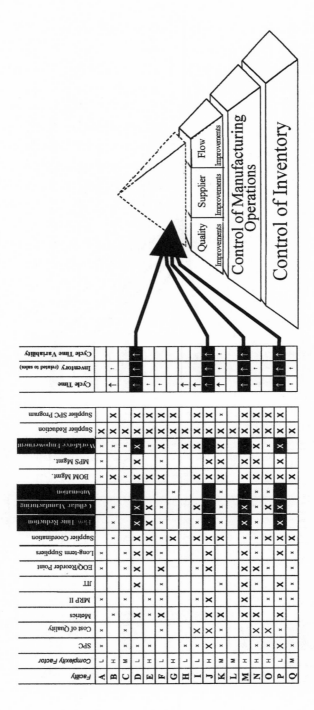

FIGURE A.7 Effect of improvements to production flow.

Much as was seen with Supplier Improvements, the tools and practices making up the enablers listed as Quality Improvements and Improvements to Production Flow only correlate to strong results when implemented in conjunction with the two foundation enablers. While these were not *essential* to improvement (strong improvements were seen by facilities D, J, M, and P in many cases without such initiatives as SPC, flow time reduction, and even cellular manufacturing), we found that their strong implementation greatly enhanced these results. This is further discussed in Chapter 6.

As shown in Figure A.6, SPC was solidly implemented in only two facilities; in each case, we see strong performance against all three of our metrics. Similarly, three of the four facilities that show strong performance in reducing cycle time, inventories, and cycle time variation have implemented strong cellular manufacturing programs (Figure A.7).

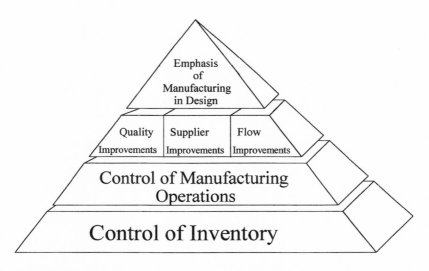

FIGURE A.8 The framework for manufacturing affordability improvement.

As was noted in Chapter 3, Emphasis of Manufacturing in Design was not included in this comparability matrix because of the lack of factory-wide examples at the time of this study. As was explained in this chapter, the optimal application of the tools and practices of this enabler was determined to depend on those already discussed. Thus, it is placed at the top of the pyramid, as depicted in Figure A.8.

Notes

Introduction

1. Womack, J. P., Jones, D. T., and Roos, D., *The Machine That Changed the World*, New York: Rawson Associates, Macmillan Publishing Co., 1990, pages 12–15 present and characterize the term "lean production."
2. *The Manufacturing Affordability Development Program: A Structured Approach to Rapidly Improved Affordability*, Final Report, Washington, DC: The Naval Air Systems Command and The Joint Strike Fighter Program Office, July 1996, page 48.

Chapter 1. An Opportunity for Advancement

1. Anderson, John D., Jr., *Introduction to Flight*, New York: McGraw-Hill Book Co., 1985, pages 239–243, provides both a historical and a technical account of the stepping-stones that ultimately led to the first manned supersonic flight.

2. *Nova* broadcast transcript #2412, "Faster than Sound," produced by WGBH-TV, Boston; Public Broadcasting System (PBS) airdate: October 14, 1997. This 50th anniversary documentary interviews a number of key players who overcame challenges leading up to supersonic flight. The progressive contributions by the Bell, NACA, and Air Force team subsequently led to success on October 14, 1947.

3. *Nova* #2412, "Faster than Sound."

4. Norman R. Augustine demonstrated this in his book *Augustine's Laws*, New York: Viking Penguin, 1986.

5. Tactical aircraft flyaway cost and year of initial operation are data that are widely available from a number of sources such as: Augustine, Norman R., *Augustine's Laws*, New York: Viking Penguin, 1986, page 109 (Figure 19) and page 110 (Figure 20). These data points were derived from his work, and our projection is only intended to show a general trend. The projection for the next-generation aircraft assumes that the present trend remains the same and uses a rough estimate as to when the next-generation aircraft might possibly enter service. Mr. Augustine reached a similar conclusion and projects the trend line further into the future to show that there could be even more dramatic consequences should this trend continue unabated throughout the next century.

6. *The Manufacturing Affordability Development Program*, Final Report, page 6. This report highlights the conclusion that direct labor represents a relatively small portion of total product cost based on data collected at over a dozen aircraft industry sites.

7. The cost of carrying inventory is expressed as an annual percentage. It is the cost of capital (money) that is associated with the value of that inventory, the quantity on hand, and the length of time the inven-

tory is carried in stock or on the manufacturing line as work in process. It can also include the cost of warehouses, personnel, and equipment used to store and track that inventory.

8. Traveled work, also called out-of-station work, is that activity which, for one reason or another, cannot be completed at the assembly station for which it was planned. As a work-around, the assembly is permitted to continue moving down the assembly line from station to station, and the tardy part is allowed to catch up at a subsequent assembly station. *Note:* The $40 million aircraft cost was arbitrarily selected for illustrative purposes; however, it is not out of line with modern prices.

9. While this percentage varied widely from facility to facility, an assumption of 20% would not be unreasonable based on what we found.

10. Contrary to popular belief, Orville and Wilbur's work represents the fruition of a century's worth of prior aeronautical development. From the invention of the practical aircraft design by George Cayley (an Englishman) in 1799, the concept of fixed-wing flight by William Samuel Henson, the practical demonstrations of John Stringfellow, the glider flights of Otto Lilienthal, and the undeniable influence of Octave Chanute on the Wrights, it is clear that the Wrights' first flight was the result of the evolutionary work of many individuals—each contributing to the successful outcome. This is covered in detail by each of the works of Drs. John D. Anderson Jr. (op. cit.) and Barnes W. McCormick *Aerodynamics, Aeronautics, and Flight Mechanics*, New York: John Wiley & Sons, 1979.

11. "The X-Planes: 50 Years of High Desert 'Right Stuff,'" NASA HQ Press Release 96-204, NASA Headquarters, Washington, D.C., October 10, 1996. This press release highlights several historical facts surrounding the first

manned supersonic flight. Among them it cites President Truman's emphasis of "the partnership that existed among the three groups [Bell, NACA, and the Air Force]."

12. While the names of many of the companies we visited have subsequently changed, these factories are still in place.

Chapter 2. The Evolution of Manufacturing in the Aircraft Industry

1. Glines, Carroll, *The Wright Brothers, Pioneers of Flight*, New York: Franklin Watts Co., 1968, page 35 notes the Wright Brothers' first engine was fabricated using only three power tools: a lathe, a drill, and a band saw.

2. We draw this conclusion by the aggregate of historical literature we cite throughout this chapter and many other examples that we do not cite directly. We have compared and contrasted this information with the techniques used for managing the aircraft industry today as witnessed during our study of the industry. The results of this study are detailed in the *Manufacturing Affordability Development Program* Final Report.

3. Lewis, Howard T., and Livesey, Charles A., *Materials Management, A Problem of the Airframe Industry*, Boston: Graduate School of Business Administration, George F. Baker Foundation, Harvard University, Volume XXXI, No. 2, July 1944 (p. 4) state that some of the techniques of mass production were attempted in the World War II time frame with varying degrees of success. The intervention and borrowing of mass production techniques from the auto industry in World War II never proved "wholly successful"—even in Britain as mentioned by John Carlton Ward Jr. in *Aircraft Production in Britain During Total War*, an address

delivered at the Ritz-Carlton in New York City on January 20, 1943 (Fairchild Engine and Airplane Corporation, pages 3–4).

4. Womack, J. P., Jones, D. T., and Roos, D., *The Machine That Changed the World*, New York: Rawson Associates, Macmillan Publishing Co., 1990, pages 12–13 and 24 present and characterize the term "craft production."

5. Simonson G. R., *The History of the American Aircraft Industry, An Anthology*, Cambridge, MA: MIT Press, 1968, page 136 states that both airframes and engines were assembled on one spot on the factory floor prior to 1940. MacKinnon, Hector Donald, Jr., *Aircraft Production: Planning and Control*, New York/Chicago: Pittman Publishing Corp., 1943, on pages 157–158 also recalls that aircraft were originally assembled in a single jig.

6. Biddle, Wayne, *Barons of the Sky: From Early Flight to Strategic Warfare, The Story of the American Aerospace Industry*, New York: Simon & Schuster, 1991, page 20, states: "Ten years after the Wrights' epochal success, flying was still an activity for visionaries and fools, with far more of the latter willing to climb aboard such a rickety deathtrap."

7. "Slipping the surly bond of earth" comes from the first line of the poem "High Flight" by John Gillespie Magee Jr., a young RAF pilot killed during World War II who composed the poem shortly before his death. It has since been published several times and is included in the preface of the book by Barnes W. McCormick *Aerodynamics, Aeronautics, and Flight Mechanics*, New York: John Wiley & Sons, 1979, page vii.

8. Rodgers, Eugene, *Flying High, The Story of Boeing and the Rise of the Jetliner Industry*, New York: Atlantic Monthly Press, 1996, page 29 describes the state of the industry just prior to World War I (with this industry annual output statistic for the year 1916).

9. Lewis and Livesey (p. 4) point out that aircraft manufacturing prior to World War II was conducted on a specific contract basis. Bowers, Peter M., *Curtiss Aircraft 1907–1947*, London: Putnam, 1979, and Mixter, Col. George Webber, *United States Army Aircraft Production Facts (WWI)*, Washington: Government Printing Office, January 1919, explain that while no real commercial market existed, the industry was largely supported by orders from the British and U.S. military.

10. Simonson (pp. 135–139) states, "Standardization was seen as a prerequisite to mass production and in order to maintain flexibility for low volume work, the aircraft industry never saw the necessity for the rigidity that mass production and standardization would bring." Standardization, or lack thereof, was a significant concern for those trying to develop production approaches to meet the expected wartime surge as discussed by Theodore Paul Wright in *Aircraft Production and National Defense*, (America in a World at War, No. 10), New York: Farrar & Rinehart, 1941, pages 13–14.

11. This plight was noted by several authors of the day, including industry executives and army representatives: Williams, G. M. (senior vice president of Curtiss Wright Corporation), *Prospects and Problems in Aviation: Growth of the Aircraft Industry*, Chicago: Chicago Association of Commerce, July 1945, pages 2–3, and Wilson, E. E. (vice chairman, United Aircraft Corporation, and president, Aeronautical Chamber of Commerce of America, Inc.), *Prospects and Problems in Aviation: Future of the Aircraft Industry*, Chicago: Chicago Association of Commerce, July 1945, pages 124–125; Mixter (p. 5); and Wright (p. 4). It should be noted that the British were in a similar situation as noted by Harald Penrose in *British Aviation: The Great War and Armistice, 1915–1919*, New York: Funk & Wagnalls, 1969, page 22,

where he describes the British industry as little more than a "collection of sheds." He explains that the French had developed more organized factories, but raw material shortages meant the French had trouble meeting demand as well. Overall World War I production totals can be found in Daniel Todd and Janice Simpson's *The World Aircraft Industry*, Dover, MA: Auburn House Publishing, 1986, pages 26–33. Simonson on page 119 states that by the time World War II approached the industry knew that a "revolution in the technology of production was needed. A shift from the job shop method to assembly line techniques was the result."

12. Rodgers (pp. 36–39); Simonson (pp. 75–78); both authors describe the boost in aviation interest worldwide as a result of Lindbergh's historic, well-publicized flight as well as the effects of the Kelly Air Mail Act of 1925, which turned over airmail routes to private industry.

13. Rae, John Bell, *Climb to Greatness: The American Aircraft Industry, 1920–1960*, Cambridge, MA: MIT Press, 1968 (pp. 39–57) devotes an entire chapter to describing Wall Street's increasing interest in the aircraft manufacturing industry, which lent legitimacy to the once-fledgling industry.

14. Rae (pp. 58–59) terms this period in history as the time of "The Airframe Revolution." It was an era where advancements in materials and manufacturing technology led to unprecedented advances in aircraft design, construction, and performance.

15. Williams (pp. 3–4) presents an account of the industry's tremendous expansion. Woods, George Bryant, *The Aircraft Manufacturing Industry, Present and Future Concepts*, New York: White, Weld & Co., 1946, page 2, gives a number of growth statistics of the overall industry at this time, including the expansion of manpower, floor space, annual production totals, and sales.

16. Woods (p. 2) points out this fact but concedes that if looked at in terms of total aircraft weight, the automobile industry's role was greater than this statistic portrays. Sir Alec Cairncross in *Planning in Wartime: Aircraft Production in Britain, Germany and the United States*, London: Macmillan, 1991 (pp. 174–177) and Rae (pp. 128–138, 157–161) explain this further by pointing out that automobile plants did virtually no aircraft assembly work, save the B-24 Willow Run plant. Instead, the auto industry had a far greater role in engine and propeller production since these tasks translated far more easily to the mass production techniques to which the automobile plants were accustomed.

17. Williams (p. 4) notes, "Production Planning had to set up simple, easily taught assignments which green hands could manage."

18. Hammer, Michael, and Champy, James, *Reengineering the Corporation: A Manifesto for Business Revolution*, New York: HarperBusiness, 1993 (p. 14) present this account.

19. MacKinnon (pp. 156–157) explains this relationship, noting that the departure from single assembly jigs to multiple subassemblies would increase efficiencies and shorten overall aircraft production span times. While James R. Yeakley and Harold C. Martin in *Aircraft Tooling Practices*, New York: Pitman Publishing Corp., 1944, on pages 4–5 point out typical examples of subassembly work in an aircraft factory, Simonson (p. 132) describes the broader implications of this approach on production planning.

20. Vander Meulen, Jacob, *Building the B-29*, Washington & London: Smithsonian Institution Press, 1995 (page 26 [BDV consortium], pages 47–51 [multiline system]). The genesis of the multiline assembly line was during prewar production of the B-17, which was produced by Boeing, Douglas and Vega, (Collison, Thomas, *Flying Fortress*,

The Story of the Boeing B-17 Bomber, New York: Charles Scribner's Sons, 1943, page 59). While we just as easily could have highlighted the B-17 manufacturing approach, the B-29 really epitomized many of the features many other lines were incorporating across the country. Rae (p. 145 [BDV]), in fact, credits Consolidated's San Diego division, working with Ford production planners, for developing the first straight-line assembly system at the B-24 Willow Run plant where work was pushed back upstream to subassembly shops. We could also have highlighted the first powered, moving assembly line for airframes used by Vultee's Downy plant (Rae, p. 127) or any of the assembly lines taking hold across the industry at that time to meet the wartime surge in demand. But the B-29 represents the culmination of the sharing of these techniques across the industry in one of the last significant "new start" aircraft programs of the war and as such was specifically designed to take advantage of these new manufacturing approaches.

21. Vander Meulen (p. 47).
22. MacKinnon (p. 101) describes how the production planning system works, and while the system differs across the industry, the procedures are fundamentally the same. Rae (p. 145) confirms this cross-pollination by stating, "This process was repeated throughout the industry so thoroughly that every major firm was making someone else's planes in addition to its own." Douglas Aircraft Company, *Engineering Planning and Scheduling in Nine Southern California Plants*, Los Angeles: Douglas Aircraft Co., 1944, describes a formal benchmarking effort of several of the engineering and production processes employed by a large cross section of the industry—further demonstrating this cross-pollination of techniques and procedures.

23. Hammer and Champy (p. 14).

24. MacKinnon (pp. 23–24) describes the 1940s formation of the organization known as "shop contact" for the purpose of bridging the communications of the shop floor and designers. He states, "With the increase in production, the project engineer [designer] could not possibly take care of the thousand-and-one details requiring engineering disposition each day."

25. MacKinnon (pp. vi, 12, 55) tells of new personnel in these departments and how they will learn how to "coordinate procedures and systems" as well as learning the "flow of authority through the factory." Lewis and Livesey (pp. 6–26) detail the creation of the materials management function and describe both its tasks as well as its interface with the remainder of the company organizations for many factories across the industry. Simonson (p. 138) describes the complexity of the manufacturing system, which requires close coordination to survive the inevitable part shortages and the end-of-the-month delivery crunch.

26. Lewis and Livesey (pp. 1, 4–5, 10, 16–17, 21, 30) describe the prevalence of hoarding of key materials as well as unpredictable shop usage and substitution of materials, which ultimately led to inaccurate inventory records. Cairncross (pp. 163–164, 167) describes various "waste and muddles" including "phantom shortages" brought on by material hoarding and overordering. Simonson (p. 138) alludes to sporadic production brought on by part shortages eventually leading to a "mad rush" to get aircraft delivered on time.

27. Hammer and Champy (p. 14).

28. Vander Meulen (pp. 47–51) presents this rationale for factory layout for the B-29 aircraft. Collison (p. 59) presents a similar argument for the B-17 aircraft.

29. Vander Meulen (pp. 32–35) describes this situation on the B-29 program. Rae (pp. 148–149) describes the prevalence of this method of operation across the industry at the time, reaching its peak in the last days of the war.

30. MacKinnon (p. 55).

31. MacKinnon (pp. 56, 85), Lewis and Livesey (pp. 4, 10, 17, 21, 22, 30–32, 42), and Cairncross (p. 168) all discuss the prevalence of this stockpiling activity at various companies throughout the industry to ensure having on hand the necessary materials when they were needed.

32. MacKinnon (pp. 40–42, 91, 208–209, 227, 230–232, 240–242) identifies the origin and operation of expediters in the aircraft industry during the early phases of World War II. Williams (pp. 4–5) discusses the evolution of the task that expediters came to provide during wartime.

33. Lewis and Livesey (p. 36) mention the crucial role that follow-up played during the war for finding and obtaining the necessary materials to keep the plant operating as smoothly as possible. Simonson (p. 138) tells of how follow-up was an essential part of supporting the end-of-the-month schedule delivery pressure.

34. Woods (p. 9) notes that "the war caused the necessary creation of certain extravagances and inefficiencies" and expressed fears that the aircraft industry might have lost its ability to be cost conscious. He also describes what he perceives to be a reluctance by some in the industry to make this transition back to more efficient operations. Lewis and Livesey (pp. 17, 22, 36, 41) describe the use of inventory as a security blanket for maintaining schedule as well as the continuing use of expediters at the end of the war. In addition, this

report also notes a certain "management ambivalence" to change.

35. Lewis and Livesey (pp. 1, 17).

36. Lewis and Livesey (pp. 2–3, 47–48) caution against the industry's new emphasis on advanced design and manufacturing technologies as the preferred approach for regaining prewar efficiencies in the postwar market. They warn against what they viewed as the industry's neglect for the improvement of inventory control, and advocate continued vigilance toward minimizing material cost and improving supplier management in the post–World War II period.

37. Rodgers (pp. 290–293) notes that during the war years the government was interested in output at all costs. He contends this behavior continued after the war since the government still remained the industry's largest customer despite relatively significant growth in the postwar commercial market.

38. Rae (Chap. 9); Simonson (p. 181) discusses the industry consolidation that occurred following the war. During this time, the commercial aircraft industry was growing despite military cutbacks. Woods (pp. 4–10) notes that industries such as automotive that entered the aircraft business during the war were satisfied with returning to their primary products to meet the enormous postwar pent-up demand. Todd and Simpson (p. 33) provide production levels for this period, showing that they did not drop off as dramatically as anticipated. So the dire consequences of the postwar drawdown predicted by Wilson and the Aircraft Industries Association of America never came to pass [Wilson (pp. 126–128) and Aircraft Industries Association of America, *The Aircraft Industry Prepares for the Future*, testimony presented by representatives of the Aero-

nautical Chamber of Commerce of America to the War Contracts Subcommittee of the Committee on Military Affairs, U.S. Senate, Washington: Aeronautical Chamber of Commerce of America, Inc., September 1944].

39. Rodgers (pp. 290–291) explains how cost consciousness did not return after the war and customers were still demanding the new aircraft in a short lead time for delivery, expediting both design and manufacture, despite the adverse impact this demand had on cost efficiency.

40. Steiner, John E., "Problems and Challenges: A Path to the Future," paper presented to the Royal Aeronautical Society, London, England, October 10, 1974, page 1.

41. Steiner (p. 4) found that the company's inventory records were highly inaccurate and this prompted a wall-to-wall physical audit of the inventory on hand. The company also attacked other causes of part shortages that hampered production flow such as convincing engine suppliers to cut their order-to-delivery lead time.

42. Rodgers (pp. 293, 303–305) explains how close Boeing came to declaring bankruptcy and how the efforts it made to regain efficiency and shed cost burdens enabled the company to regain its financial footing.

43. Rodgers (p. 292) notes that part shortages continued for some time. Steiner (p. 4) concedes that a lack of coordination with customers and suppliers when internal factory cycle times were cut caused more traveled work and higher costs. Supplied items could not keep pace, and customers were unable to take early delivery of the finished aircraft thus costing Boeing $2 million per month in interest and taxes on the undelivered inventory.

44. Womack, Jones, and Roos (p. 48) present these facts on production quantities and other constraints facing the Toyota Motor Company.

45. Womack, Jones, and Roos (pp. 54, 56–57) detail the history of Toyota's strengthened cooperation with its workforce. They state that Toyota agreed to give workers lifetime employment but in return expected them to be flexible in work assignments and actively participate in helping the company eradicate waste and improve its production processes.

46. Ohno, Taiichi, *The Toyota Production System: Beyond Large Scale Production*, Portland, OR: Productivity Press, 1988 (pp. 1, 7, 10, 39) explains the unique market requirements that Toyota had to fill: The company had to cut costs while providing a wide diversity of products at low volume in order to survive in the postwar Japanese economy.

47. Ohno (pp. 38–39, 41) explains that this continuous improvement approach is necessary to ensure the smooth, continuous operation of the Toyota production line. Shingo, Shigeo, *A Study of the Toyota Production System: From an Industrial Engineering Viewpoint*, Portland, OR: Productivity Press, 1989 (pp. 14–15, 117–119) notes the Toyota production system focus on root cause correction of defects as well as the necessity for a 100% defect-free product. Womack, Jones, and Roos (p. 149) describe the purpose of *kaizen* and its role in continuously improving the production process at Toyota.

48. Womack, Jones, and Roos (pp. 60, 61) note that Toyota integrated its close-knit supplier base into design and manufacturing decisions and would even lend workers to the suppliers to deal with workload surges.

49. Taylor, Alex, III, "How Toyota Defies Gravity," *Fortune*, Vol. 136, No. 11, December 8, 1997 (pp. 100–108 points out that Toyota is the world leader in product develop-

ment. He describes how Toyota uses the design process to gain even further efficiencies on the factory floor. Womack, Jones, and Roos (pp. 97, 111–119) explain that because of Toyota's excellent product development process, new model introduction isn't as much of a problem for Toyota as it is for others. They contend that Toyota's factory productivity recovers from new model introduction within four months. For American plants it takes five months for the same level of recovery; it takes a full year for the European factory.

50. National Center for Manufacturing Sciences (NCMS), *Competing in World-Class Manufacturing: America's 21st Century Challenge*, Homewood, IL: Business One Irwin, 1990 (pp. 22, 29, 41, 54, 56, 89, 203, 217, 241) notes that Ford, GM, and Chrysler were put on the defense in the early 1980s by the superior quality coming out of Japan at the time. Each of them took steps, such as Ford's "Quality Is Job 1" program, to try to "equal Japan on quality." By 1985, 90% of the Fortune 500 companies were using things such as Quality Circles that were borrowed from Japan.

51. Harmon, Roy L., *Reinventing the Factory II: Managing the World Class Factory*, New York: Free Press, 1992 (p. 11) expresses his concern about the "false understanding" being brought back from fact-finding trips to Japan claiming the key of Japan's success lies in tools such as Quality Circles and Statistical Process Control alone. Schonberger, Richard J., *World Class Manufacturing*, New York: Free Press, 1986 (pp. 13, 33) speaks of the Zero Defects programs in the United States as well as the lackluster results of Quality Circle implementation in America.

52. Wilson (pp. 126–128) predicted significantly stunted growth and financial turmoil in the industry after the war based on similar upheaval that followed World War

I. Woods (pp. 4–10) describes how the actual downsizing was accomplished with considerably less pain than was predicted.

Chapter 3. A Study in Manufacturing Affordability

1. *The Manufacturing Affordability Development Program*, Final Report (v, 43, 48): Throughout this book we refer to a study known as the Manufacturing Affordability Development Program (MADP). The MADP was sponsored by the Joing Advanced Strike Technology (JAST) program (whose name later changed to the Joint Strike Fighter program) beginning in 1995 and culminated in a well-attended Industry Day debriefing of the results in August 1996. The Final Report was released to the public and can be downloaded from the World Wide Web at: www. jast.mil. Shortly thereafter, an article summarizing the MADP findings was published (Fulghum, David A., "JSF Manufacturing Combed for 25% Cut," *Aviation Week & Space Technology*, July 7, 1997, p. 49).

2. On December 16, 1998, the Garrity Tool Company won honorable mention for the 1998 Indianapolis Mayor's Eagle Award in the categories of Innovative Methods/Research and Development, Commitment/Vision for the Future of Indianapolis and Workforce Development, and won the award for Environmental Recycling efforts. On October 25, 1999, Garrity Tool was presented with the 1999 State of Indiana Quality Improvement Award in recognition of its outstanding leadership and achievement after being nominated by one of its customers that had recently selected Garrity Tool Company as a preferred supplier.

3. As stated in *The Manufacturing Affordability Development Program*, Final Report, page 41, these numbers represent a composite of the best performance from all of the facilities reviewed.

4. This graphic is not intended as a precise representation; it should merely be taken as a general description of a trend.

5. *The Manufacturing Affordability Development Program*, Final Report.

6. For cost estimation purposes, aircraft are frequently broken down into their basic subcomponents: airframe, engine, and avionics. The *Manufacturing Affordability Development Program*, Final Report (page 42) estimates the contribution of each of these systems on the total production cost of a modern tactical aircraft. By adding together validated product savings from the best airframe, engine, and avionics producers, the total "roll-up" of these savings was shown to be as much as 25% of the production cost of the completed tactical aircraft. However, only after specific savings were verified—both their association with the types of lean activities reviewed in this study and their magnitude—were they used to form this aggregation. We concluded that this yielded a conservative estimate, since it excludes many other improvements we witnessed but could not quantify.

Chapter 4. The Principle of Variation Management

1. National Center for Manufacturing Sciences (NCMS) (pp. 57, 64) The authors state that the contribution of Statistical Process Control to overall "quality progress" in Japan reached an all-time high in 1980 and then began to decline as Design of Experiments replaced it as the most significant contributor to quality progress until 1990.

Chapter 5. Production and Inventory Control: The Cornerstone Enablers

1. In addition to our own observations, this is acknowledged in Smith, S. B., *Computer Based Production and Inventory Control*, Englewood Cliffs, NJ: Prentice Hall, Inc., 1989, pages 23–25.

2. The Economic Order Quantity figure is based on charts we found throughout the industry. The construction of this type of chart is covered in Figure 17.4 of Vollman, Thomas E., Berry, William L., and Whybark, D. Clay, *Manufacturing Planning and Control Systems*, New York: Irwin Professional Publishing, 1992, p. 708.

3. The Reorder Point figure is constructed based on descriptions provided during our visits. This figure was constructed primarily to depict how variation is absorbed by safety stocks. We found this approach to be generally described by Figures 17.3 and 17.7 of Vollman, Berry, and Whybark, pages 706 and 714, respectively.

Chapter 6. The Impact of Production Flow and Quality

1. Cells: "The Celling Out of America," *Economist*, Vol. 333, No. 7894, December 17, 1994 (pp. 63–64; NCMS (pp. 13, 87, 310, 323); Schonberger (*World Class Manufacturing*, pp. 10–11) describes the growing use of manufacturing cells in the United States. Worker empowerment: NCMS (pp. 40–41, 61) and Schonberger (p. 37) discuss the increasing use of "employee involvement programs." SPC: The increasing role of SPC and its use in both the United States and Japan are discussed in NCMS (39–40, 51–53, 64, 117). Flow reductions: The increasing use of flow reduction techniques is covered by several articles: Taylor, Alex, III, "How Toyota Defies Gravity," *Fortune*, Vol. 136, No. 11, December 8, 1997 (pp. 100–108; "The Keys to Success," *Industry Week*, Vol. 246, No. 16, September 1, 1997 (pp. 20–21); Moskal, Brian, "Best Plants: General Motors," *Industry Week*, Vol, 244, No. 19, October 16, 1995 (pp. 33–34); Sheridan, John H., "*Kaizen* Blitz," Industry Week, Vol. 246, No. 16, September 1, 1997 (pp. 18–27).

2. "The Kindergarten That Will Change the World," *Economist*, Vol. 334, No. 7904, March 4, 1995 (p. 63) points out the sharp contrast of the low-automation approach being applied to the RAV4 compared with previous reports that Toyota was aiming for a highly automated factory making extensive use of robots and other high-technology approaches.

3. Schonberger (pp. 63–64) presents and characterizes the term "pre-automation."

4. Shingo, Shigeo, *A Study of the Toyota Production System: From an Industrial Engineering Viewpoint*, Productivity Press: Portland, OR, 1989 (p. 21) defines the term *poka-yoke* as: "Successive, self, and source inspection can all be achieved through the use of poka-yoke methods. Poka-yoke achieves 100 percent inspection through mechanical or physical control." He goes on to describe the various types and means of implementing this method. Ohno (p. 6) uses the term *baka-yoke* in his book to describe a "foolproofing system to prevent defective products." But from what we have seen, the term *poka-yoke* seems to be pervasive today.

5. NCMS (p. 238) refers to the creation of "Total Productive Maintenance" by GE in the 1950s. Most other texts and what we have witnessed around the industry suggests a modification to the term "Total Preventive Maintenance" as being more descriptive of the proactive nature to which this method has evolved.

6. This is a characterization that is presented by Schonberger (p. 111).

7. Harmon (pp. 169–174) makes the assertion that the grouping approach was employed by the Japanese originally as a means of design simplification and standardization.

8. Lewis, Howard T., and Livesey, Charles A., *Materials Management, A Problem of the Airframe Industry*, Boston:

Graduate School of Business Administration, George F. Baker Foundation, Harvard University, Volume XXXI, No. 2, July, 1944 (pp. 33, 44, 46) note that a couple of companies in the aircraft industry had experimented with grouping parts by either their common features or the predictability of their demand for inventory control purposes and to ensure a smoother work flow. Womack, James P., and Jones, Daniel T., *Lean Thinking*, New York: Simon & Schuster, 1996 (pp. 156 and 164) describe Pratt & Whitney's early approach of aligning production operations to take advantage of the natural flow of work. They further discuss factors during World War II that may have led the company toward mass production practices.

9. Schonberger, Richard, *Japanese Manufacturing Techniques (JMT)*, New York: Free Press, 1982 (p. 115) cites Burbridge, 1975. The first known publications we have seen with regard to group technologies were English translations of the works of Russian author Mitrofanov: S. P. Mitrofanov, *Scientific Principle of Group Technology* (1959), translated by G. Harris, edited by T. J. Grayson, London: National Lending Library for Science and Technology, 1966, and subsequent translation of Mitrofanov's 1970 work, *Scientific Organization of Batch Production*, by the U.S. Air Force as part of the Integrated Computer Aided Manufacturing (ICAM) project in 1976.

10. Harmon (pp. 172–174) describes the use of the concept he witnessed at Yamaha.

11. Schonberger (*World Class Manufacturing*, p.112) presents this finding.

12. Harmon (pp. 172–174) describes the approach taken in design which later led to actions on the assembly floor to simplify and reorganize operations and take advantage of the standard grouping to do mixed production.

13. Goldratt, Eliyahu M. *The Goal*, Great Barrington, MA: North River Press, 1992, provides a comprehensive discussion on the benefits and methods of seeking out and addressing factory bottlenecks.

Chapter 7. Improved Supplier Responsiveness

1. *The Manufacturing Affordability Development Program*, Final Report, page 28 provides this number.
2. Palmeri, Christopher, "Mix and Match the Technology," *Forbes*, Vol. 160, No. 10, November 3, 1997 (p. 186) presents these facts about Wal-Mart and EDI.
3. Gansler, Jacques S., *The Defense Industry*, Cambridge, MA: MIT Press, 1980, (p. 138) discusses how small defense subcontractors have high pricing pressures while subjected to high risk.
4. Gansler (p. 131) notes that in 1974 numerous suppliers left the defense industry [which is closely tied to the aerospace industry] for more traditional commercial sector work and didn't come back despite increased demand.
5. Cramer, Harold, *The Elements of Probability Theory*, New York: John Wiley & Sons, 1955 (p. 79) presents these statistical derivations which define how normal distributions can be summed. In addition, assuming that each distribution is equal and centered about the same mean, he describes how this relationship can be simplified to the second equation we present in this text.

Chapter 8. The Role of Product Design

1. Taylor (pp. 100–108); Schonberger (*World Class Manufacturing*, pp. 144–147); Funke, Christopher C. (Boeing Co.), "Concurrent Engineering in the Aircraft Industry," *Aerospace Engineering*, September 1997, (pp. 15–19); Buchholz, Kami, "Concurrent Engineering Offers Ripple Benefits," *Aerospace Engineering*, November 1996;

NCMS (pp. 120, 207, 208) all describe the use of concurrent design being implemented in a number of different industries.

2. Viemeister, Peter, *Start All Over: An American Experience: People, Places and Lessons Learned*, Bedford, VA: Hamilton's, 1995 (p. 187).

3. MacKinnon (pp. 23–24).

4. Variability Simulation Analysis (VSA) is a registered trademark of the Variation Simulation Analysis Company. This company markets a piece of software that was initially used by the auto industry to study manufacturing tolerance buildup between major assemblies and body panels. Some companies in the aircraft industry have adopted the use of this tool to simulate the effects of variation and tolerance buildup of their CAD (computer-aided design)-based designs.

5. Smith, Spencer B., *Computer Based Production and Inventory Control*, Englewood Cliffs, NJ: Prentice Hall, 1989 (p. 299).

6. We are thankful to Dr. James Womack, president of the Lean Enterprise Institute, for pointing out to us that Toyota does not rely on learning curves. When reviewing accounts of the details of Toyota production system such as the definitive work by Monden, Yasuhiro, *The Toyota Production System*, 3d ed., Norcross, GA: Engineering and Management Press, 1998, we noted that learning curves are conspicuous by their absence. It appears that Toyota has steered away from them in favor of a more precise approach. According to Monden and Shingo, Shigeo, *A Study of the Toyota Production System from an Industrial Engineering Standpoint*, Portland, OR: Productivity Press, 1989, Toyota designs its manufacturing processes in concert with the design of the product—taking care to remove sources of waste prior to producing the product. After developing precise time standards for each element

of work performed at a given workstation, Toyota managers can closely track any deviations. Further factory refinements resulting from *kaizen* events are put into place in preplanned increments to maintain factory balance. Thus, Toyota does not appear to informally "learn" how to improve the process with each successive unit. It seems that learning curves would only serve to introduce cycle time creep in a system that is precisely balanced on predetermined cycle times.

Chapter 9. The Culture of Manufacturing

1. Rader, James, *Penetrating the U.S. Automobile Market, German and Japanese Strategies, 1965–1976*, Ann Arbor, MI: UMI Research Press, 1980 (p. 2) quotes an industry study of the time that suspended concerns of foreign competition by referring to it as "insignificant." NCMS (pp. 4–5) provides a brief history of the conditions occurring in the auto industry at that time. Foreign competition was largely ignored by a General Motors report, "The Automotive Industry: A Case Study of Competition," Detroit: General Motors Corporation, October 1968 (p. 22).

2. Schonberger, Richard, *Japanese Manufacturing Techniques (JMT)*, New York: Free Press, 1982 (p. 200) cites the Harbour Report of 1981 which compared the cost of producing an automobile in the United States and Japan, highlighting the quality and productivity gap. The Massachusetts Institute of Technology's International Motor Vehicle Program, headed by Jim Womack, Dan Jones, and Dan Roos also conducted extensive research into this area during the mid-1980s, which was discussed in their book *The Machine That Changed the World*.

3. Shingo (p. 147) discusses how Toyota leverages the flexibility of its workforce, enforces strict rules of the Toyota

production system, provides workers with standard training followed by a high level of autonomy, and achieves
high productivity as a result.

4. Abernathy, William, *The Productivity Dilemma*, Baltimore:
   Johns Hopkins University, 1978 (pp. 3, 4, 8, 9, 60, 171,
   172) discusses his theory for declining productivity (that
   it has an inverse relationship with advances in technology) in the U.S. auto industry. Hashimoto, Masanori,
   *Japanese Labor Market in a Comparative Perspective with the
   U.S.*, Kalamazoo, MI: W. E. Upjohn Institute for Employment Research, 1990, compares the condition and statistics of the labor markets of Japan and the United States.
   Rader looks at the various factors, such as product dump-
   ing, affecting foreign penetration of the U.S. auto market.
   Harmon (p. ix) cites high cost and inferior quality, with
   deeper causes being high inventory and longer lead times;
   NCMS (Ref. 1 of Chapter 1: Office of Technology Assessment, "Making Things Better—Competing in Manufacturing," 1989, which also highlighted this productivity
   gap).

5. Marsh, Robert M., and Mannari, Hiroshi, *Modernization
   and the Japanese Factory*, Princeton, NJ: Princeton University Press, 1976, along with other observations, contend
   that the Japanese success is dependent on their unique
   culture.

6. NCMS (p. 162) mentions General Electric's "sluggish implementation" of Just-in-Time techniques. Harmon (p.
   21) argues the failure of many of these Japanese "fads" and
   disputes the common belief of American management
   that the tools would only work in the Japanese culture as
   the reason for failure. Schonberger (*JMT*: pp. 3, 15,
   24–25) echoes the argument that tools such as JIT are not
   solely dependent on the Japanese culture and economy in
   order to be made to work.

7. Harmon (p. 21) draws this conclusion based on America's

growing number of successes at implementing these tools. Schonberger (*WCM*: p. ix) mentions the collective works of a number of authors who have all concluded that the use of these techniques is not limited to the Japanese culture.

8. NCMS (p. 310) highlights GM's implementation of manufacturing cells. Taylor (p. 2) mentions how GM, Ford, and Chrysler have all "borrowed bits and pieces" of the Toyota production system. Moskal (p. 2) discusses the successes that the GM Delphi plant has had with a pull manufacturing system with kanban cards.

9. *The Manufacturing Affordability Development Program*, Final Report, page 47.

Chapter 10. Changing the Paradigm

1. Ohno (pp. xiv, 97, 100, 103, 107, 108) describes his appreciation for Ford's approaches and a general admiration of Ford's genius. He also mentions the inspiration Ford provides for Ohno to develop his own techniques that suit the conditions of Toyota.

2. Lacey, Robert, *Ford: The Men and the Machine*, Boston: Little, Brown and Co., 1986 (page 120) mentions this criticism of Ford, which appeared in the *Wall Street Journal* at the time of Ford's announcement.

3. Gelderman, Carol, *Henry Ford: The Wayward Capitalist*, New York: Dial Press, 1981 (p. 52) describes this meeting and Ford's approach to convincing his board members.

4. Gelderman (p. 51); Collier, Peter, and Horowitz, David, *The Fords: An American Epic*, New York: Summit Books, 1987 (p. 65); and Lacey (p. 121) all list this problem as one of the causes for Ford's forthcoming drastic action.

5. Ford, Henry, *Today and Tomorrow*, Portland, OR: Productivity Press, 1988 (pp. 158–159); Gelderman (p. 54);

Lacey (pp. 117, 120, 129). A stable workforce performing repetitive tasks learns how to become more efficient with every repetition until it reaches a solid level of productivity. Ford realized that continued employee turnover would prevent his workforce from reaching this level of efficiency. Instead, new employees would have to be brought in and taught the system, and would need time to acclimate to the new system before beginning their search for the most productive methods. This perpetual problem could not be tolerated if Ford was to realize his goal of running the most productive automobile factory.

6. Gelderman (p. 52) describes the board's resistance to Ford's proposed wage increase.

7. Lacey (pp. 117, 129) explains that Ford got more production while the employees worked fewer hours. Gelderman (p. 54), mentions that this gave Ford discretion to pick the very best workers for his line. Ford also received higher output from his workers for less money per auto. Collier (pp. 66–67) notes that as a result of the change the workforce was much more stable, and absenteeism dropped by more than an order of magnitude.

8. Lacey (pp. 117, 120) notes that Ford's change resulted in increased throughput and profit. Gelderman (p. 54) mentions that the wage increase was the cause of the record profits the company began to receive. Ford (p. 159) explains his rationale for the eight-hour day. He states that the company found that eight hours of work is the optimal amount of time for receiving peak performance from his workers on a consistent basis.

9. Lacey (p. 120) lists the cost of the change as less than $10 million for the first year, but the shareholders still voted themselves an $11.2 million dividend that same year—indicating success. Gelderman (p. 54) also confirms that the Five Dollar Day was responsible for the significant increase in Ford's profits.

10. Ohno (pp. 5, 105–107, 115) tells of his admiration for Ford's ability to develop "brilliant inverse conceptions." He draws a similarity between himself and Ford by stating the he, too, tries whenever possible to form inverse conceptions. So convinced is he of this powerful approach that he also encourages his workers to think this way.

11. Ohno (pp. 35–36).

12. Gelderman (p. 51) presents these figures, which demonstrate Ford's cost-cutting successes.

13. Shingo, Shigeo, *A Study of the Toyota Production System: From an Industrial Engineering Viewpoint*, Productivity Press: Portland, OR, 1989 (p. 139) makes this statement. Dr. Shingo is the coinventor of the Just-in-Time system. He is cast in the light of being the academic or teacher whereas Mr. Ohno was the manager. Both of them express concern over the misapplication or piecemeal use of the tools of the Toyota production system.

14. Ohno (pp. 38–39, 44, 95–96) describes production leveling as a critical, indispensable first step in the implementation of the Toyota production system. Unfortunately, this is a step which we have seen many omit when trying to imitate this production system and its techniques.

Chapter 11. A Final Note . . .

1. Ohno (pp. 10–13, 36–39, 44) explains the first steps he used when implementing the Toyota production system: establishing the production flow to synchronize the factory's operations and production leveling to smooth the flow of products through the facility.

2. This analogy is similar to one put forth by Dr. Shingo when cautioning against the use of temporary fixes to relieve the symptoms rather than addressing the root cause of the problem. He states on page 74 of *A Study of the*

Toyota Production System, "These solutions are like using an ice pack to cure appendicitis—it may relieve the pain for a while, but only an appendectomy will prevent recurrence. This is the Toyota approach—to discover and implement solutions that permanently prevent a problem from recurring."

Appendix

1. The Appendix of this book and the figures used therein are taken directly from or based largely on the *Manufacturing Affordability Development Program*, Final Report.

Index